# Moto Guzzi 750, 850 and 1000 V-twins Owners Workshop Manual

## by Mansur Darlington

**Models covered:**
750 S 748 cc. UK August 1974 to December 1974
750 S3 748 cc. UK May 1975 to November 1976
850 T 844 cc. USA 1975 to 1977
850 T 844 cc. UK August 1974 to November 1976
850 T3 844 cc. UK & USA 1975 to 1978
Le-Mans 844 cc. UK June 1976 to 1978
V-1000 Convert 949 cc. UK & USA 1975 to 1978

ISBN 978 0 85696 339 1

*(339-7U8)*

**Haynes Group Limited**
**Haynes North America, Inc**

www.haynes.com

# Acknowledgements

Our thanks are due to Moto Guzzi SpA of Mandello del Lario, Como, Italy who encouraged the publishers to prepare this manual. Our thanks also go to Moto Guzzi Concessionaires, Luton, Bedfordshire, who supplied the 850T3 which features in the photographs, and to Mr P.H. Slinn, Service Manager of that Company for his assistance and advice.

We are indebted to Ted Hutchins of E.A. Taylor, Misterton, Somerset, for numerous tips gleaned from his experience with the marque, and to Mervyn Bleach who allowed us to photograph his V-1000 I-Convert.

Brian Horsfall gave necessary assistance with the overhaul and devised ingenious methods for overcoming the lack of service tools. Les Brazier took the photographs that accompany the text. Jeff Clew edited the text.

Finally, we would also like to acknowledge the help of the Avon Rubber Company who kindly supplied illustrations and advice about tyre fitting, NGK Spark Plugs (UK) Ltd for information about spark plug maintenance and electrode conditions, and Contact Developments of Reading who allowed us to reproduce Dell 'Orto carburettor illustrations, and who supplied us with the originals so swiftly.

# About this manual

The author learnt his motorcycle mechanics by trial and error - possibly more by error. In presenting this manual it is hoped that these errors can be avoided by others. Only by supervising the work himself, under conditions similar to those in which the amateur mechanic works, can the author ensure that the text is a true and concise record of procedure. Thus in the photographs, the hands shown are those of the author.

The machine selected had covered an average mileage, so that any problems encountered would be typical of those facing the average owner.

Moto Guzzi service tools were not used, as generally an alternative method of removing or replacing a part was possible. Certain special tools would simplify work on this unusual (for motorcycles) design. They may be hired by callers at some Moto Guzzi agents. A torque wrench should be begged or borrowed for use where torque wrench settings are given. Some car accessory shops and tool hire companies will often loan one.

Always have all tools and replacement parts to hand before commencing work. Baking trays or similar containers are useful for putting small parts in. Replace nuts and washers on the studs they fitted where possible, this avoids loss. Unless otherwise mentioned, reassembly should be carried out in reverse order to dismantling.

Each of the seven Chapters is divided into numbered Sections. Within the Sections are numbered paragraphs. Cross-reference throughout this manual is quite straightforward and logical. When reference is made, 'See Section 6.10', it means Section 6, paragraph 10 in the same Chapter. If another Chapter were meant it would say, 'See Chapter 2, Section 6.10'.

All photographs are captioned with a Section/paragraph number to which they refer, and are always relevant to the Chapter text adjacent.

Figure numbers (usually line illustrations) appear in numerical order, within a given Chapter. 'Fig. 1.1' therefore refers to the first figure in Chapter 1.

Left-hand and right-hand descriptions of the machines and their components refer to the left and right of a given machine when normally seated.

Motorcycle manufacturers continually make changes to specifications and recommendations, and these, when notified, are incorporated into our manuals at the earliest opportunity.

We take great pride in the accuracy of information given in this manual, but motorcycle manufacturers make alterations and design changes during the production run of a particular motorcycle of which they do not inform us. No liability can be accepted by the authors or publishers for loss, damage or injury caused by any errors in, or omissions from, the information given.

# Contents

Note: General description and specifications are given in each Chapter immediately after the list of Contents. Fault diagnosis is given at the end of each appropriate Chapter

Left and right-hand view of Moto Guzzi 850T3 California

# Introduction to the Moto Guzzi 750, 850 and 1000 V-twins

Moto Guzzi SpA was founded in 1921 by Carlo Guzzi and Giorgio Parodi. With Guzzi as the engineer, the first production model was a 500 cc horizontal engined single cylinder machine; an engine configuration which, although unusual, was to be one of the Moto Guzzi hallmarks until after the Second World War. The 500 cc engine was modified continually in the light of successful racing experience, and in common with the 250 cc engine which soon supplemented it, was available in IOE, OHV and OHC form at various times during its production run. By 1934, when a 175 cc machine was produced, all models could be supplied either in roadgoing or racing trim, with or without a spring frame.

The early post-war era saw a completely new range of models to be built concurrently with the later models of the single cylinder horizontal engine type. The range included machines with both two-stroke and four-stroke engines of between 63 cc and 235 cc capacity. Most of these models outlived the earlier singles, the production of which ceased in the late '50's. Moto Guzzi, in addition to models which though successful as racing machines were ostensibly produced as road models, have a history of producing one-off machines designed solely for the purpose of winning races in the hands of their own riders. These include the 250 cc supercharged model, the 500 cc 120° V-twin, and, after the War, the in-line DOHC water-cooled four and the even more magnificent 500 cc V-8.

The Moto Guzzi transverse V-twin of today, of which the V-1000 I-Convert is the latest addition, utilizes an engine designed originally for use in a light four-wheel transporter manufactured for the armed forces. The compactness of the engine and the transverse configuration has made the engine ideally suited as a motorcycle power unit. Included in the range currently in production is the 850 Le Mans model, a sporting road machine which with little modification can be used as a production racer. This model has superseded the 750S and S3 models, which were developments of the earlier V7 sport. The touring field is catered for by the 850T and the T3, the latter model incorporating an integral front and rear braking system. The 850T3 California is the out and out American style tourer. The V-1000 I-Convert model is considered by many to have the ideal specifications for a long-distance tourer, having weather equipment and panniers similar to those of the 850 California, a 1000 cc engine and an almost unique transmission system utilizing a torque converter and a two-speed gearbox. The success of the transverse V-twin is reflected in the new range of lightweight Moto Guzzi models of 350 cc and 500 cc which share a purpose-built engine unit similar to the earlier larger machines, and retaining the shaft final drive.

# Dimensions and weight

|  | 750S | 750S3 | 850 Le Mans | 850T3 and V-1000 |
|---|---|---|---|---|
| **Overall length** | 2165 cm (85.2 in) | 2165 cm (85.2 in) | 2190 cm (86.2 in) | 2200 cm (86.5 in) |
| **Overall width** | 700 cm (27.5 in) | 680 cm (26.8 in) | 720 cm (28.3 in) | 780 cm (34.0 in) |
| **Overall height** | 1035 cm (40.7 in) | 1020 cm (40 in) | 1030 cm (40.5 in) | 1060 cm (46.0 in) |
| **Wheel base** | 1470 cm (58 in) | 1470 cm (58 in) | 1470 cm (58.0 in) | 1470 cm (58.0 in) |
| **Weight** | 225 kg (495 lbs) | 230 kg (507 lbs) | 198 kg (437 lbs) | 240 kg (560 in) |

# Ordering spare parts

Every Moto Guzzi authorised dealer undertakes to stock those genuine Moto Guzzi parts that are required frequently. Other parts which are required less frequently can be obtained by the dealer from an area stockist, or from the importers. Only genuine, approved Moto Guzzi parts must be used.

When ordering parts, quote the full frame and engine numbers. If painted parts are required, include the colour.

The frame number is stamped on the right-hand side of the steering head, and on a plate on the front of the steering head.

The engine number is on the left of the crankcase, above the oil-level dipstick boss.

Pattern parts may sometimes be available at lower cost, but they do not necessarily make a satisfactory replacement for the originals. There are cases where reduced life or sudden failure has occurred, to the overall detriment of performance, and perhaps safety.

Some of the more expendable items such as spark plugs, bulbs, tyres, oil and grease etc, can be obtained from accessory shops and motor factors. They have convenient opening hours, charge lower prices, and can often be found not far from home. It is also possible to obtain parts by mail order from specialists who advertise in the motorcycle magazines.

Engine number location

Frame number location

# Safety first!

Professional motor mechanics are trained in safe working procedures. However enthusiastic you may be about getting on with the job in hand, do take the time to ensure that your safety is not put at risk. A moment's lack of attention can result in an accident, as can failure to observe certain elementary precautions.

There will always be new ways of having accidents, and the following points do not pretend to be a comprehensive list of all dangers; they are intended rather to make you aware of the risks and to encourage a safety-conscious approach to all work you carry out on your vehicle.

### Essential DOs and DON'Ts

**DON'T** start the engine without first ascertaining that the transmission is in neutral.

**DON'T** suddenly remove the filler cap from a hot cooling system – cover it with a cloth and release the pressure gradually first, or you may get scalded by escaping coolant.

**DON'T** attempt to drain oil until you are sure it has cooled sufficiently to avoid scalding you.

**DON'T** grasp any part of the engine, exhaust or silencer without first ascertaining that it is sufficiently cool to avoid burning you.

**DON'T** allow brake fluid or antifreeze to contact the machine's paintwork or plastic components.

**DON'T** syphon toxic liquids such as fuel, brake fluid or antifreeze by mouth, or allow them to remain on your skin.

**DON'T** inhale dust – it may be injurious to health (see *Asbestos* heading).

**DON'T** allow any spilt oil or grease to remain on the floor – wipe it up straight away, before someone slips on it.

**DON'T** use ill-fitting spanners or other tools which may slip and cause injury.

**DON'T** attempt to lift a heavy component which may be beyond your capability – get assistance.

**DON'T** rush to finish a job, or take unverified short cuts.

**DON'T** allow children or animals in or around an unattended vehicle.

**DON'T** inflate a tyre to a pressure above the recommended maximum. Apart from overstressing the carcase and wheel rim, in extreme cases the tyre may blow off forcibly.

**DO** ensure that the machine is supported securely at all times. This is especially important when the machine is blocked up to aid wheel or fork removal.

**DO** take care when attempting to slacken a stubborn nut or bolt. It is generally better to pull on a spanner, rather than push, so that if slippage occurs you fall away from the machine rather than on to it.

**DO** wear eye protection when using power tools such as drill, sander, bench grinder etc.

**DO** use a barrier cream on your hands prior to undertaking dirty jobs – it will protect your skin from infection as well as making the dirt easier to remove afterwards; but make sure your hands aren't left slippery. Note that long-term contact with used engine oil can be a health hazard.

**DO** keep loose clothing (cuffs, tie etc) and long hair well out of the way of moving mechanical parts.

**DO** remove rings, wristwatch etc, before working on the vehicle – especially the electrical system.

**DO** keep your work area tidy – it is only too easy to fall over articles left lying around.

**DO** exercise caution when compressing springs for removal or installation. Ensure that the tension is applied and released in a controlled manner, using suitable tools which preclude the possibility of the spring escaping violently.

**DO** ensure that any lifting tackle used has a safe working load rating adequate for the job.

**DO** get someone to check periodically that all is well, when working alone on the vehicle.

**DO** carry out work in a logical sequence and check that everything is correctly assembled and tightened afterwards.

**DO** remember that your vehicle's safety affects that of yourself and others. If in doubt on any point, get specialist advice.

**IF,** in spite of following these precautions, you are unfortunate enough to injure yourself, seek medical attention as soon as possible.

### Asbestos

Certain friction, insulating, sealing, and other products – such as brake linings, clutch linings, gaskets, etc – contain asbestos. *Extreme care must be taken to avoid inhalation of dust from such products since it is hazardous to health.* If in doubt, assume that they *do* contain asbestos.

### Fire

Remember at all times that petrol (gasoline) is highly flammable. Never smoke, or have any kind of naked flame around, when working on the vehicle. But the risk does not end there – a spark caused by an electrical short-circuit, by two metal surfaces contacting each other, by careless use of tools, or even by static electricity built up in your body under certain conditions, can ignite petrol vapour, which in a confined space is highly explosive.

Always disconnect the battery earth (ground) terminal before working on any part of the fuel or electrical system, and never risk spilling fuel on to a hot engine or exhaust.

It is recommended that a fire extinguisher of a type suitable for fuel and electrical fires is kept handy in the garage or workplace at all times. Never try to extinguish a fuel or electrical fire with water.

**Note:** *Any reference to a 'torch' appearing in this manual should always be taken to mean a hand-held battery-operated electric lamp or flashlight. It does **not** mean a welding/gas torch or blowlamp.*

### Fumes

Certain fumes are highly toxic and can quickly cause unconsciousness and even death if inhaled to any extent. Petrol (gasoline) vapour comes into this category, as do the vapours from certain solvents such as trichloroethylene. Any draining or pouring of such volatile fluids should be done in a well ventilated area.

When using cleaning fluids and solvents, read the instructions carefully. Never use materials from unmarked containers – they may give off poisonous vapours.

Never run the engine of a motor vehicle in an enclosed space such as a garage. Exhaust fumes contain carbon monoxide which is extremely poisonous; if you need to run the engine, always do so in the open air or at least have the rear of the vehicle outside the workplace.

### The battery

Never cause a spark, or allow a naked light, near the vehicle's battery. It will normally be giving off a certain amount of hydrogen gas, which is highly explosive.

Always disconnect the battery earth (ground) terminal before working on the fuel or electrical systems.

If possible, loosen the filler plugs or cover when charging the battery from an external source. Do not charge at an excessive rate or the battery may burst.

Take care when topping up and when carrying the battery. The acid electrolyte, even when diluted, is very corrosive and should not be allowed to contact the eyes or skin.

If you ever need to prepare electrolyte yourself, always add the acid slowly to the water, and never the other way round. Protect against splashes by wearing rubber gloves and goggles.

### Mains electricity and electrical equipment

When using an electric power tool, inspection light etc, always ensure that the appliance is correctly connected to its plug and that, where necessary, it is properly earthed (grounded). Do not use such appliances in damp conditions and, again, beware of creating a spark or applying excessive heat in the vicinity of fuel or fuel vapour. Also ensure that the appliances meet the relevant national safety standards.

### Ignition HT voltage

A severe electric shock can result from touching certain parts of the ignition system, such as the HT leads, when the engine is running or being cranked, particularly if components are damp or the insulation is defective. Where an electronic ignition system is fitted, the HT voltage is much higher and could prove fatal.

# Routine maintenance

Periodic routine maintenance is a continuous process that commences immediately the machine is used and continues until the machine is no longer fit for service. It must be carried out at specified mileage recordings or on a calendar basis if the machine is not used regularly, whichever is the soonest. Maintenance should be regarded as an insurance policy, to help keep the machine in the peak of condition and to ensure long, trouble-free service. It has the additional benefit of giving early warning of any faults that may develop and will act as a safety check, to the obvious advantage of both rider and machine alike.

The various maintenance tasks are described under their respective mileage and calendar headings. Accompanying photos or diagrams are provided, where necessary. It should be remembered that the interval between the various maintenance tasks serves only as a guide. As the machine gets older, is driven hard, or is used under particularly adverse conditions, it is advisable to reduce the period between each check.

For ease of reference each service operation is described in detail under the relevant heading. However, if further general information is required it can be found within the manual in the relevant Chapter.

Although no special tools are required for routine maintenance, a good selection of general workshop tools are essential. Included in the tools must be a range of metric ring or combination spanners, a selection of crosshead screwdrivers, and two pairs of circlip pliers, one external opening and the other internal opening.

---

**Weekly or every 300 miles**

---

### Tyre pressures
1   Check the tyre pressures with a pressure gauge that is known to be accurate. Always check the pressures when the tyres are cold. If the tyres are checked after the machine has travelled a number of miles, the tyres will have become hotter and consequently the pressure will have increased, possibly as much as 8 psi. A false reading will therefore always result.

| Model | Front tyre | Rear tyre |
|---|---|---|
| 750S, S3, 850T3 and Le Mans | 29 psi (2 kg-cm$^2$) | 33 psi (2.3 kg-cm$^2$) |
| 850T | 26 psi (1.8 kg-cm$^2$) | 33 psi (2.3 kg-cm$^2$) |
| V-1000 | 30 psi (2.1 kg-cm$^2$) | 34 psi (2.4 kg-cm$^2$) |

When carrying a pillion passenger the rear tyre pressure should be increased by 3 psi (0.2 kg-cm$^2$). When travelling at continuous high speeds an additional 3 psi (0.2 kg-cm$^2$) should be added to both front and rear tyres.

### Engine oil
2   With the machine positioned upright, on its centre stand, run the engine for a few minutes then allow the oil level to settle. The combined filler plug and dipstick is situated in the left-hand crankcase and should be fully screwed in when the oil level measurement is taken. Never run the engine with the oil level lower than the lower level mark. Avoid overfilling as this causes higher crankcase pressures which may damage the oil seals. Replenish with SAE 10W/50 or 20W/50 engine oil.

### Torque converter reservoir oil level - V-1000 models
3   Remove the left-hand frame side cover by pulling the lower edge from the rubber securing bush and lifting the cover off the upper hooks. By means of the dipstick in the filler cap check the torque converter hydraulic fluid level. The level should come between the upper and lower marks. Replenish if necessary with Dexron ® Automatic Transmission Fluid. **DO NOT under any circumstances use engine oil or hydraulic brake fluid.**

### Safety inspection
4   Give the whole machine a close visual inspection, checking for loose nuts and fittings, frayed control cables and damaged brake hoses etc.

### Legal inspection
5   Check that the lights, horn and flashing indicators function correctly, also the speedometer.

Check and if necessary replenish engine oil

Check torque converter fluid level by means of dipstick

---

**Monthly or every 500 miles**

Complete the tasks listed under the weekly/300 mile heading and then carry out the following checks.

*Tyre damage*
1  Rotate each wheel and check for damage to the tyres, especially splitting on the sidewalls. Remove any stones or other objects caught between the treads. This is particularly important on the front tyre, where rapid tyre deflation due to penetration of the inner tube will almost certainly cause total loss of control of the machine.

*Spoke tension*
2  Check the spokes for tension, by gently tapping each one with a metal object. A loose spoke is identifiable by the low pitch noise emitted when struck. If any one spoke needs considerable tightening, it will be necessary to remove the tyre and inner tube in order to file down the protruding spoke end. This will prevent it from chafing through the rim band and piercing the inner tube.
On machines fitted with cast alloy spoked wheels a close visual inspection is necessary, checking for cracks and similar structural damage.

*Rear brake adjustment - 750S and 850T models only*
3  When the rear brake is in correct adjustment the total brake pedal travel measured at the toe tread should be within the range 0.8 - 1.2 in (20 - 30 mm). If the travel is greater or less than this carry out the necessary adjustment by means of the shouldered nut at the brake arm end of the control rod.
The actual brake pedal travel is really a matter of choice, but should not be so tight that the brake linings bind on the drum. Conversely, excess brake pedal travel will prevent quick operation of the brake.

*Battery electrolyte level*
4  Release the seat catch and lift up the dualseat. Remove the tool tray and detach the battery retaining strap. Remove the filler plug or plugs and check the battery electrolyte level. The solution should just cover the battery plates. If required, replenish using distilled water. Do not fill to a level more than 5 mm (3/16 in) over the top of the plates.

---

**Two monthly or every 2,000 miles**

Complete the checks listed under the weekly 300 mile and monthly 500 mile headings and then carry out the following tasks:

*Engine oil change*
1  The oil should be changed regularly at the prescribed intervals, more particularly in the case of 750S and most 850T models, where no in-line oil filter cartridge is used. Drain the oil into a container of more than 3.5 litres (7.3/6.0 US/Imp pints). The drain plug is located in the rear wall of the sump. Oil drainage will be accelerated and improved if the engine has reached normal working temperature; the oil will be thinner and so flow more readily. Refit the drain plug and replenish the engine through the filler orifice with approximately 3 litres of SAE 10W/50 or 20W/50 engine oil. Check the level with the dipstick and then pour in a further amount of oil until the level reaches the maximum mark.

*Gearbox oil level*
2  Remove the gearbox level plug from the right-hand side of the machine. The oil level should be just below that of the lower threads in the aperture. If required, replenish with a MP SAE 90 or EP 90 gearbox oil through the filler orifice.

*Rear bevel drive box level*
3  Remove the bevel box level plug and check the oil level. If necessary, replenish to a level just below the filler orifice with MP SAE 90 or EP 90 gearbox oil.

*Tappet adjustment*
4  Remove the spark plugs and detach each rocker cover after removing the retaining screws. A small amount of oil will spill out of each rocker chamber when the covers are removed. Adjustment of the clearance between the rocker arms and valve stems should be made with the engine **COLD**. Rotate the engine until one piston is at top dead centre (TDC) on the compression stroke (both valves closed). Check the rocker clearances with a 0.22 mm (0.0085 in) feeler gauge. Adjustment is made by loosening the locknut on the rocker arm and screwing the adjuster inwards or outwards, as necessary. Hold the adjuster firmly, tighten the locknut and recheck. The feeler gauge should be a light sliding fit.
Repeat the operation on the other cylinder.

Check gearbox oil level and ...

... refill if necessary with correct oil

Rear bevel box oil level plug

Adjust tappet clearance with engine COLD

Complete all the checks listed in the previous schedules and then complete the following:

### Hydraulic brake fluid level

1   Check the level of the brake fluid in both the front and rear master cylinder reservoirs. Before removing the front brake reservoir cap place the handlebars in such a position that the reservoir is as upright as possible; this will prevent spillage. If fluid spills onto paintwork or plastic fittings, wash it off immediately. Hydraulic fluid is a good paintstripper.

Replenish if the level is below the lower face of the diaphragm, with the diaphragm positioned correctly. Never allow the fluid to fall more than 6 mm below the maximum mark which is measured at the waist of the diaphragm. Use hydraulic fluid of DOT 3 (USA) or SAE J1703 specification. Never use engine oil or similar fluids. If the level of fluid in either of the reservoirs is excessively low, check the pads for wear. If the pads are not worn, suspect a fluid leakage in the system. This must be rectified immediately.

V-1000 and Le Mans models are fitted with a float-operated switch in the rear master cylinder, which illuminates a warning light if the fluid falls below a pre-set level. This does not, however, preclude the necessity of regular fluid level checks.

### Air filter cleaning

2   This service item does not apply to Le Mans models. Raise the dualseat and detach the battery retaining straps and the petrol tank retaining strap. Disconnect the negative battery lead followed by the positive lead and lift the battery from position. After detaching the petrol feed pipes at the tap unions, lift the tank up at the rear and away from the machine.

Because of the design of the air filter rubber duct, which connects the air filter to the carburettors, the carburettors must be detached at the cylinder heads. Unscrew the three socket screws holding each inlet stub to the cylinder head and swing both carburettors away, after pulling them from the rubber ducting.

Unscrew the single nut from the rod projecting through the front of the air filter box and withdraw the breather box. The hose to the box may be pulled off. Lift out the air filter element.

The element is of the corrugated dry paper type. Knock the loose dust from the element and blow out the ingrained dirt from the inside of the filter, using an air jet. If the element is badly soiled, perforated or oil soaked, it should be renewed without question. Poor performance and an over-rich mixture will result from a blocked filter.

When refitting the filter assembly, note that the free plate at the front of the breather box must be located correctly with the projection on the box.

### Spark plug cleaning

3   Remove the spark plugs and clean them, using a wire brush. Clean the electrodes with fine emery paper or cloth and then reset the gaps to 0.6 mm (0.023 in) with the correct feeler gauge. Before replacing the plugs, smear the threads with a small amount of graphite grease, to aid future removal. Where high speed CW275 L plugs are being used, the correct gap is 0.5 mm (0.019 in).

### Cleaning and adjusting the contact breakers

4   In order to inspect and adjust the contact breakers, the housing cap, retained by two screws, must be removed. To aid access for easy removal and subsequent attention, the rear of the petrol tank should be raised a few inches after detaching the retaining strap and the petrol pipes. Support the tank on a bunch of rags or a wooden block.

Rotate the engine until one set of points is open and examine the contact faces. Slight irregularities in the faces may be removed using a fine swiss file or a strip of emery paper backed by a piece of tin. If they are dirty, pitted or burnt, it will be necessary to remove them for further attention, as described in Section 3 of Chapter 4.

Repeat the process on the second set of points.

The correct contact breaker gap, when the points are in the fully open position, is within the range 0.42 - 0.48 mm (0.016 - 0.018 in) for 850T models and 0.37 - 0.43 mm (0.014 - 0.017 in) for all other models. Adjustment is effected by slackening the two screws passing through the contact breaker fixed point plate and using a screwdriver inserted in the notch provided, moving the fixed contact near to or further away from the moving contact. Ensure that the points are in the fully open position when this adjustment is made or a false setting will result. Tighten the two screws and recheck the gap; the feeler gauge should be a light sliding fit between the faces.

Repeat the process on the other set of contact points. Before refitting the housing cap, clean the points faces using a clean rag dipped in methylated spirits. This will ensure that the points are perfectly clean and prevent the faces picking-up prematurely. Apply a few drops of thin oil to the cam lubricator wick. Do not overlubricate or the excess oil may find its way to the points, causing ignition failure.

*Ignition timing*

5   It is important that the ignition timing is checked regularly

and accurately. Check the timing as described in Chapter 4, Section 7.

---

**Six monthly or every 6,000 miles**

---

Carry out the tasks described in the weekly, monthly, two monthly and four monthly sections and then attend to the following:

*Fuel filter cleaning*

1   Disconnect the petrol pipes at the petrol tap unions. The pipes are retained either by screw clips or spring clips. Drain the petrol tank completely by fitting temporarily suitable lengths of hose to the taps. Unscrew each tap by applying a spanner to the hexagonal nut above the tap body. A gauze filter is fitted to each tap. Remove the screw passing through each pipe union at the carburettors. Pull the unions away and displace the circular filter screens.

Clean the screens and tap filters in petrol, removing stubborn deposits with a soft brush. If the gauze is perforated, the tap or screen should be renewed. Refit the components, ensuring that they are not overtightened. A little petrol resistant sealing compound applied to the tap threads will help prevent leakage.

Do not allow fluid level to drop in either reservoir

Carburettors must be removed to gain access to element

Clean cylindrical tap filters and ...

... the filter screen at each carburettor union

## Changing the gearbox and bevel box oil

2   Drain the contents of the gearbox and the bevel box after the machine has been on a run of sufficient length to allow the oil to reach normal working temperature. The gearbox drain plug on the V-1000 models is fitted to the lower edge of the end cover. On all other models the drain plug is in the casing base.

Replace both drain plugs and refill the gearbox with 0.6 ltr (1.26/1.05 US/Imp pts) for V-1000 models and 0.751 ltr (1.75/1.33 US/Imp pts) for all other models.

The rear bevel box on all models has a capacity of 0.250 ltr (8.4/7.04 US/Imp fl oz). A mixture of MP SAE 90 or EP 90 gear oil should be made up to this amount with 20 cc (¾ oz) of Molykote type A or a molybdenum disulphide transmission additive.

## Battery connections

3   Disconnect the negative battery lead followed by the positive lead. Clean the terminals and the leads thoroughly with wire wool or emery paper. Refit and tighten the leads. Apply a coating of petroleum jelly to the two terminals covering all the exposed metal, to prevent further corrosion.

## Air filter

4   The air filter element should be renewed at this service interval regardless of its condition. Removal is described in the Section on filter cleaning under the 4,000 mile service heading.

## Spark plugs

5   It is recommended that the spark plugs are renewed at approximately 6,000 mile intervals. Although a spark plug may give good service after this mileage peak efficiency will have been lost. Refer to Chapter 4 specifications for the correct plug type.

## Nine monthly or every 9,000 miles

Complete all the previously listed tasks and then carry out the following operation:

## Oil filter cleaning

1   At 9,000 miles approximately, or at every fifth oil change, the oil filter screen should be removed and cleaned and the oil filter element, where fitted, renewed. After draining the oil in the normal way undo the sump retaining screws and lift the sump away. If care is taken, the sump gasket can often be reused. Unscrew the oil filter cartridge and discard it. Remove the centre screw from the oil filter screen, after bending down the ear of the locking plate. Displace the screen and wash it thoroughly in petrol. Allow the screen to dry before refitting. Fit a new filter cartridge and replace the sump. During this service operation the sump itself should be cleaned thoroughly.

## Yearly or every 12,000 miles

The yearly maintenance schedule constitutes a minor overhaul and in addition to all the preceding maintenance tasks the following should be carried out:
1   Check the condition of the rear brake linings as described in Chapter 6, Section 13.
2   Remove, inspect and regrease the wheel bearings. Refer to Chapter 6, Section 10.
3   Relubricate the steering head bearings. See Chapter 5, Section 5.

## Front fork oil change

4   The lubricating oil in each front fork leg should be drained and replenished. Place the machine on the centre stand and remove the socket screw from the centre of each fork top bolt. Place blocks below the engine so that the front wheel is clear of the ground and remove the two top bolts. Unscrew the drain plug at the lower rear of each fork leg and allow the oil to drain. Replace the drain plugs and refill each fork leg with the correct quantity of Dexron ® Automatic Transmission Fluid or equivalent.

Quantity per leg:

| Model | Fork leg capacity (each) |
|---|---|
| 750S and 850T | 50 cc (1.7/1.4 US/Imp fl oz) |
| 850T3 | 60 cc (2.0/1.7 US/Imp fl oz) |
| 750S3 and V-1000 | 70 cc (2.7/2.0 US/Imp fl oz) |
| Le Mans | 120 cc (4.0/3.4 US/Imp fl oz) |

After filling, refit the top bolts, lower the front wheel onto the ground and fit the damper retaining socket bolts.

## Additional routine maintenance

### 1   Brake pads: examination and replacement

The rate of brake pad wear is dependent on the conditions under which the machine operates, weight carried and the style of riding, consequently it is difficult to advise on specific inspection intervals. Whatever inspection interval is chosen, bear in mind that the rate of wear will not be constant.

Prise the fluted cover off the top of each caliper and check the width of each pad. If any pad has worn to less than 6 mm (0.2362 in) both pads in that set must be renewed.

The pads may be removed without detaching either the wheel or caliper, as follows:

Depress one end of the pad pin retaining spring and withdraw the freed pin. Remove the spring and the second long pin and then lift the tapered pin out of position in the pads. The pads may be displaced one at a time. Slot the new set of pads into place, if necessary pushing the pistons back into the caliper halves to gain the added clearance. Refit the pins by reversing the dismantling procedure.

### 2   Clutch adjustment

In common with brake pad wear, clutch wear and the resultant necessary adjustment depends on operating conditions and the style of riding. Adjust the clutch, when necessary, as follows:

Rotate the adjuster wheel on the handlebar control until the free play measured between the lever and lever stock is approximately 4 mm (5/32 in). There must always be some free play in the cable or the clutch pushrod will bind, causing premature wear. In addition, the clutch plates may slip due to the slight amount of lift imparted. If the required latitude of adjustment is not available at the handlebar control lever, use the adjuster at the lower end of the cable, after loosening the locknut.

### 3   Control cable lubrication

Use motor oil or an all-purpose oil to lubricate the control cables. A good method for lubricating the cables is shown in the accompanying illustration, using a plasticine funnel. This method has a disadvantage in that the cables usually need removing from the machine. An hydraulic cable oiler which pressurises the lubricant, overcomes this problem. Do not lubricate nylon lined cables as the oil will cause the nylon to swell, thereby causing total cable seizure.

Remove and discard old oil filter

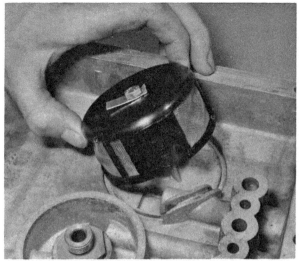

Detach and clean oil filter screen

Prise off cap to inspect brake pads

Withdraw all pins and spring to ...

... enable removal of pad sets

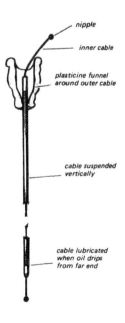

nipple

inner cable

plasticine funnel
around outer cable

cable suspended
vertically

cable lubricated
when oil drips
from far end

Oiling control cable

# Working conditions and tools

When a major overhaul is contemplated, it is important that a clean, well-lit working space is available, equipped with a workbench and vice, and with space for laying out or storing the dismantled assemblies in an orderly manner where they are unlikely to be disturbed. The use of a good workshop will give the satisfaction of work done in comfort and without haste, where there is little chance of the machine being dismantled and reassembled in anything other than clean surroundings. Unfortunately, these ideal working conditions are not always practicable and under these latter circumstances when improvisation is called for, extra care and time will be needed.

The other essential requirement is a comprehensive set of good quality tools. Quality is of prime importance since cheap tools will prove expensive in the long run if they slip or break when in use, causing personal injury or expensive damage to the component being worked on. A good quality tool will last a long time, and more than justify the cost.

For practically all tools, a tool factor is the best source since he will have a very comprehensive range compared with the average garage or accessory shop. Having said that, accessory shops often offer excellent quality tools at discount prices, so it pays to shop around. There are plenty of tools around at reasonable prices, but always aim to purchase items which meet the relevant national safety standards. If in doubt, seek the advice of the shop proprietor or manager before making a purchase.

The basis of any tool kit is a set of open-ended spanners, which can be used on almost any part of the machine to which there is reasonable access. A set of ring spanners makes a useful addition, since they can be used on nuts that are very tight or where access is restricted. Where the cost has to be kept within reasonable bounds, a compromise can be effected with a set of combination spanners – open-ended at one end and having a ring of the same size on the other end. Socket spanners may also be considered a good investment, a basic $3/8$ in or $1/2$ in drive kit comprising a ratchet handle and a small number of socket heads, if money is limited. Additional sockets can be purchased, as and when they are required. Provided they are slim in profile, sockets will reach nuts or bolts that are deeply recessed. When purchasing spanners of any kind, make sure the correct size standard is purchased. Almost all machines manufactured outside the UK and the USA have metric nuts and bolts, whilst those produced in Britain have BSF or BSW sizes. The standard used in USA is AF, which is also found on some of the later British machines. Others tools that should be included in the kit are a range of crosshead screwdrivers, a pair of pliers and a hammer.

When considering the purchase of tools, it should be remembered that by carrying out the work oneself, a large proportion of the normal repair cost, made up by labour charges, will be saved. The economy made on even a minor overhaul will go a long way towards the improvement of a toolkit.

In addition to the basic tool kit, certain additional tools can prove invaluable when they are close to hand, to help speed up a multitude of repetitive jobs. For example, an impact screwdriver will ease the removal of screws that have been tightened by a similar tool, during assembly, without a risk of damaging the screw heads. And, of course, it can be used again to retighten the screws, to ensure an oil or airtight seal results. Circlip pliers have their uses too, since gear pinions, shafts and similar components are frequently retained by circlips that are not too easily displaced by a screwdriver. There are two types of circlip pliers, one for internal and one for external circlips. They may also have straight or right-angled jaws.

One of the most useful of all tools is the torque wrench, a form of spanner that can be adjusted to slip when a measured amount of force is applied to any bolt or nut. Torque wrench settings are given in almost every modern workshop or service manual, where the extent to which a complex component, such as a cylinder head, can be tightened without fear of distortion or leakage. The tightening of bearing caps is yet another example. Overtightening will stretch or even break bolts, necessitating extra work to extract the broken portions.

As may be expected, the more sophisticated the machine, the greater is the number of tools likely to be required if it is to be kept in first class condition by the home mechanic. Unfortunately there are certain jobs which cannot be accomplished successfully without the correct equipment and although there is invariably a specialist who will undertake the work for a fee, the home mechanic will have to dig more deeply in his pocket for the purchase of similar equipment if he does not wish to employ the services of others. Here a word of caution is necessary, since some of these jobs are best left to the expert. Although an electrical multimeter of the AVO type will prove helpful in tracing electrical faults, in inexperienced hands it may irrevocably damage some of the electrical components if a test current is passed through them in the wrong direction. This can apply to the synchronisation of twin or multiple carburettors too, where a certain amount of expertise is needed when setting them up with vacuum gauges. These are, however, exceptions. Some instruments, such as a strobe lamp, are virtually essential when checking the timing of a machine powered by CDI ignition system. In short, do not purchase any of these special items unless you have the experience to use them correctly.

Although this manual shows how components can be removed and replaced without the use of special service tools (unless absolutely essential), it is worthwhile giving consideration to the purchase of the more commonly used tools if the machine is regarded as a long term purchase Whilst the alternative methods suggested will remove and replace parts without risk of damage, the use of the special tools recommended and sold by the manufacturer will invariably save time.

# Recommended lubricants

| Component | AGIP product | Alternative grade |
|---|---|---|
| **ENGINE** ... ... ... ... ... | SINT 2000 SAE 10W/50 | 10W/50 or 20W/50 |
| **GEARBOX** ... ... ... ... ... | F.1 Rotra MP SAE 90 | Hypoy SAE 90 |
| **TORQUE CONVERTER** ... ... ... | Dexron ® F.1 ATF | Dexron ® ATF |
| **REAR BEVEL DRIVE** ... ... ... | F.1 Rotra MP SAE 90 and Molykote Type A | Hypoy SAE 90 |
| **TELESCOPIC FORK** ... ... ... | Dexron ® F.1 ATF | Dexron ® ATF |
| **DISC BRAKES** ... ... ... ... | DOT 3 (USA) | J1703 hydraulic fluid |

Moto Guzzi recommend the use of AGIP lubricants for all applications on their machines. AGIP, however, is not available in the UK or the USA and as such a good quality brand of the equivalent correct grade may be substituted.

# Maintenance and capacities data

| | Metric | Imp | US |
|---|---|---|---|
| **Engine oil:** | | | |
| 750S, S3 and 850T ... ... ... ... ... ... ... | 3.5 ltr | 6.16 pt | 3.70 qt |
| All others ... ... ... ... ... ... ... | 3.0 ltr | 5.28 pt | 3.17 qt |
| **Gearbox oil:** | | | |
| V-1000 ... ... ... ... ... ... ... ... ... | 0.600 ltr | 1.05 pt | 1.26 pt |
| All others ... ... ... ... ... ... ... ... | 0.750 ltr | 1.33 pt | 1.75 pt |
| **Torque converter fluid** ... ... ... ... ... ... ... | 1.5-1.7 ltr | 2.6-3.0 pt | 3.17-3.6 pt |
| **Rear bevel drive oil** ... ... ... ... ... ... ... | 0.250 ltr | 7.04 fl oz | 8.4 fl oz |
| **Front fork oil:** | | | |
| 750S and 850T ... ... ... ... ... ... ... | 50 cc | 1.4 fl oz | 1.7 fl oz |
| 850T3 ... ... ... ... ... ... ... ... | 60 cc | 1.7 fl oz | 2.0 fl oz |
| 750S3 and V-1000 ... ... ... ... ... ... | 70 cc | 2.0 fl oz | 2.7 fl oz |
| Le Mans ... ... ... ... ... ... ... ... | 120 cc | 3.4 fl oz | 4.0 fl oz |
| **Tappet clearances (COLD)** ... ... ... ... ... ... | 0.22 mm (0.009 in) all valves | | |
| **Spark plug gap:** | | | |
| Standard plug ... ... ... ... ... ... | 0.6 mm (0.023 in) | | |
| High speed plug ... ... ... ... ... ... | 0.5 mm (0.019 in) | | |
| **Contact breaker gap:** | | | |
| 850T ... ... ... .. ... ... ... ... ... | 0.42-0.48 mm (0.016-0.018 in) | | |
| All other models ... ... ... ... ... ... | 0.37-0.43 mm (0.014-0.017 in) | | |

| | Front | Rear* |
|---|---|---|
| **Tyre pressures:** | | |
| 750S, S3, 850T3 and Le Mans ... ... ... ... ... | 29 psi (2.0 kg cm²) | 33 psi (2.3 kg cm²) |
| 850T ... ... ... ... ... ... ... ... ... | 26 psi (1.8 kg cm²) | 33 psi (2.3 kg cm²) |
| V-1000 ... ... ... ... ... ... ... ... | 30 psi (2.1 kg cm²) | 34 psi (2.4 kg cm²) |

*\* Add 3 psi (0.2 kg cm²) when carrying pillion passenger.*
*Add 3 psi (0.2 kg cm²) to both tyres when travelling at a continuous high speed.*

# Chapter 1 Engine

**Contents**

---

**Specifications**

## V-1000 model
### Engine

| | |
|---|---|
| Type ... ... ... ... ... ... ... ... ... | Air cooled, 90° V-twin four stroke |
| Bore ... ... ... ... ... ... ... ... ... | 88 mm (3.46 in) |
| Stroke ... ... ... ... ... ... ... ... ... | 78 mm (3.07 in) |
| Capacity ... ... ... ... ... ... ... ... | 948.8 cc (57.9 cu in) |
| Compression ratio ... ... ... ... ... ... ... | 9.2 : 1 |
| bhp (SAE) ... ... ... ... ... ... ... | 71 @ 6,500 rpm |
| Max rpm ... ... ... ... ... ... ... ... | |
| Direction of rotation ... ... ... ... ... ... | Clockwise, viewed from front |

### Valve clearances (engine cold)

| | |
|---|---|
| Inlet ... ... ... ... ... ... ... ... ... | 0.22 mm (0.009 in) |
| Exhaust ... ... ... ... ... ... ... ... ... | 0.22 mm (0.009 in) |

### Valve timing

| | |
|---|---|
| Inlet opens ... ... ... ... ... ... ... ... | 20° BTDC |
| Inlet closes ... ... ... ... ... ... ... ... | 52° ABDC |
| Exhaust opens ... ... ... ... ... ... ... | 52° BBDC |
| Exhaust closes ... ... ... ... ... ... ... | 20° ATDC |

**Note:** Timing is set with valve clearances adjusted to 0.5 mm (0.020 in)

## Valves

| | |
|---|---|
| Seat angle ... ... ... ... ... ... ... ... | 45° 30′ |
| Inlet: | |
|     head diameter ... ... ... ... ... ... ... | 40.8 - 41.0 mm (1 .606 - 1.614 in) |
|     stem diameter ... ... ... ... ... ... ... | 7.972 - 7.987 mm (0.3138 - 0.3144 in) |
| Exhaust: | |
|     head diameter ... ... ... ... ... ... ... | 35.8 - 36.0 mm (1.409 - 1.417 in) |
|     stem diameter ... ... ... ... ... ... ... | 7.965 - 7.980 mm (0.3136 - 0.3142 in) |
| Valve guide inside diameter ... ... ... ... ... ... | 8.00 - 8.022 mm (0.3149 - 0.3158 in) |
| Valve guide/stem clearance: | |
|     inlet ... ... ... ... ... ... ... ... | 0.013 - 0.050 mm (0.0005 - 0.0019 in) |
|     exhaust ... ... ... ... ... ... ... | 0.020 - 0.057 mm (0.0008 - 0.0022 in) |
| Valve guide outside diameter: | |
|     inlet ... ... ... ... ... ... ... | 14.064 - 14.075 mm (0.5537 - 0.5541 in) |
|     exhaust ... ... ... ... ... ... ... | 14.107 - 14.118 mm (0.55541 - 0.5558 in) |
| Valve guide housing inside diameter ... ... ... ... ... | 14.00 - 14.018 mm (0.5512 - 0.5519 in) |
| Valve spring free length: | |
|     outer ... ... ... ... ... ... ... ... | 52.6 mm (2.07 in) |
|     inner ... ... ... ... ... ... ... ... | 44.7 mm (1.770 in) |

## Rocker arms

| | |
|---|---|
| Bush inside diameter ... ... ... ... ... ... ... | 15.032 - 15.059 mm (0.5918 - 0.5929 in) |
| Spindle outside diameter ... ... ... ... ... ... | 14.982 - 14.994 mm (0.5899 - 0.5903 in) |
| Bush/spindle clearance ... ... ... ... ... ... ... | 0.038 - 0.076 mm (0.0015 - 0.003 in) |

## Camshaft

| | |
|---|---|
| Journal diameter: | |
|     flywheel side ... ... ... ... ... ... ... | 31.984 - 32.000 mm (1.2592 - 1.2598 in) |
|     timing side ... ... ... ... ... ... ... | 46.984 - 47.00 mm (1.849 - 1.850 in) |
| Housing diameter: | |
|     flywheel side ... ... ... ... ... ... ... | 32.025 - 32.050 mm (1.2607 - 1.2623 in) |
|     timing side ... ... ... ... ... ... ... | 47.025 - 47.050 mm (1.8511 - 1.8529 in) |
| Clearance: | |
|     flywheel side ... ... ... ... ... ... ... | 0.025 - 0.066 mm (0.001 - 0.0035 in) |
|     timing side ... ... ... ... ... ... ... | 0.025 - 0.066 mm (0.001 - 0.0035 in) |

## Tappets

| | |
|---|---|
| Guide bore diameter: | |
|     standard ... ... ... ... ... ... ... ... | 22.021 - 22.00 mm (0.8669 - 0.8661 in) |
|     1st oversize ... ... ... ... ... ... ... | 22.071 - 22.050 mm (0.8689 - 0.8681 in) |
|     2nd oversize ... ... ... ... ... ... | 22.121 - 22.100 mm (0.8709 - 0.8700 in) |
| Tappet outside diameter | |
|     standard ... ... ... ... ... ... ... | 21.996 - 21.978 mm (0.8659 - 0.8652 in) |
|     1st oversize ... ... ... ... ... ... ... | 22.046 - 22.028 mm (0.8679 - 0.8672 in) |
|     2nd oversize ... ... ... ... ... ... | 22.096 - 22.018 mm (0.8699 - 0.8668 in) |
| Clearance ... ... ... ... ... ... ... ... | 0.004 - 0.043 mm (0.0015 - 0.0016 in) |

## Crankshaft

| | |
|---|---|
| Mainbearing journal outside diameter: | |
|     flywheel side ... ... ... ... ... ... ... | 53.970 - 53.931 mm (2.1248 - 2.1233 in) |
|     timing side ... ... ... ... ... ... ... | 37.975 - 37.959 mm (1.4951 - 1.4944 in) |
| Main bearing inside diameter: | |
|     flywheel side ... ... ... ... ... ... ... | 54.000 - 54.019 mm (2.1260 - 2.1267 in) |
|     timing side ... ... ... ... ... ... ... | 38.000 - 38.016 mm (1.4961 - 1.4967 in) |
| Clearance: | |
|     flywheel side ... ... ... ... ... ... ... | 0.030 - 0.068 mm (0.0011 - 0.0027 in) |
|     timing side ... ... ... ... ... ... ... | 0.025 - 0.057 mm (0.0010 - 0.0022 in) |

**Note:** Main bearings are available in three oversizes of 0.20 mm (0.0079 in) increments

| | |
|---|---|
| Big end journal diameter: | |
|     standard class A ... ... ... ... ... ... ... | 44.008 - 44.014 mm (1.7325 - 1.7328 in) |
|     standard class B ... ... ... ... ... ... ... | 44.014 - 44.020 mm (1.7328 - 1.7330 in) |

**Note:** On original assembly the big-end journals are selectively matched with two classes of big end bearing. Class A marked white is matched with Class B crankshaft marked white. Class B bearings marked blue on the connecting rod are matched with Class A crankshaft, marked blue. When the crankshaft requires re-grinding three bearing oversizes are available.

| | |
|---|---|
| Big-end bearing | |
|     1st oversize ... ... ... ... ... ... ... ... | 0.254 mm (0.010 in) |
|     2nd oversize ... ... ... ... ... ... ... | 0.508 mm (0.020 in) |
|     3rd oversize ... ... ... ... ... ... ... | 0.762 mm (0.030 in) |

| | |
|---|---|
| Big-end bearing clearance ... ... ... ... ... ... | 0.050 - 0.085 mm (0.002 - 0.0032 in) |
| Big-end axial float ... ... ... ... ... ... | 0.030 - 0.040 mm (0.0011 - 0.0015 in) |

Small end bearing

| | |
|---|---|
| Small end bush diameter ... ... ... ... ... ... | 22.025 - 22.045 mm (0.8670 - 0.8767 in) ) |
| Gudgeon pin diameter ... ... ... ... ... ... ... | 22.00 - 22.004 mm (0.8661 - 0.8663 in) ) |
| Clearance ... ... ... ... ... ... ... ... | 0.021 - 0.045 mm (0.0008 - 0.0017 in) |

## Cylinder barrel diameter

| | |
|---|---|
| Class A ... ... ... ... ... ... ... ... ... | 88.00 - 88.009 mm (3.4645 - 3.4649 in) |
| Class B ... ... ... ... ... ... ... ... ... | 88.009 - 88.018 mm (3.4649 - 3.4652 in) |

**Note:** Cylinders are selectively matched to pistons of same class on original assembly. On subsequent rebore, two oversizes of piston are available.

## Pistons and rings

Piston outside diameter:

| | |
|---|---|
| Class A ... ... ... ... ... ... ... ... | 87.933 - 87.942 mm (3.4619 - 3.4622 in) |
| Class B ... ... ... ... ... ... ... ... | 87.942 - 87.951 mm (3.4622 - 3.4626 in) |

Piston oversize:

| | |
|---|---|
| 1st oversize ... ... ... ... ... ... ... | 0.40 mm (0.0157 in) |
| 2nd oversize ... ... ... ... ... ... ... | 0.60 mm (0.032 in) |

Ring end gap:

| | |
|---|---|
| compression rings ... ... ... ... ... ... | 0.30 - 0.45 mm (0.0118 - 0.018 in) |
| oil control ring ... ... ... ... ... ... | 0.25 - 0.40 mm (0.010 - 0.015 in) |

Side clearance:

| | |
|---|---|
| top ring ... ... ... ... ... ... ... | 0.030 - 0.062 mm (0.0011 - 0.0024 in) |
| 2nd ring ... ... ... ... ... ... ... | 0.030 - 0.062 mm (0.0011 - 0.0024 in) |
| 3rd ring ... ... ... ... ... ... ... | 0.030 - 0.062 mm (0.0011 - 0.0024 in) |
| oil control ring ... ... ... ... ... ... | 0.025 - 0.040 mm (0.0010 - 0.0015 in) |

## 850T, 850T3 and 850 California models
### Engine

| | |
|---|---|
| Type ... ... ... ... ... ... ... ... ... | Air cooled, 90° V-twin four stroke |
| Bore ... ... ... ... ... ... ... ... ... | 83 mm (3.267 in) |
| Stroke ... ... ... ... ... ... ... ... ... | 78 mm (3.070 in) |
| Capacity ... ... ... ... ... ... ... ... | 844 cc (51.49 cu in) |
| Compression ratio ... ... ... ... ... ... ... | 9.5 : 1 |
| bhp (SAE) ... ... ... ... ... ... ... ... | 68.5 @ 7,000 rpm |
| Direction of rotation ... ... ... ... ... ... ... | Clockwise, viewed from front |

### Valve clearances (engine cold)

| | |
|---|---|
| Inlet ... ... ... ... ... ... ... ... | 0.22 mm (0.009 in) |
| Exhaust ... ... ... ... ... ... ... ... | 0.22 mm (0.009 in) |

### Valve timing

| | |
|---|---|
| Inlet opens ... ... ... ... ... ... ... ... | 20° BTDC |
| Inlet closes ... ... ... ... ... ... ... ... | 52° ABDC |
| Exhaust opens ... ... ... ... ... ... ... ... | 52° BBDC |
| Exhaust closes ... ... ... ... ... ... ... ... | 20° ATDC |

**Note:** Timing is set with valve clearances adjusted to 1.5 mm (0.060 in)

### Valves

| | |
|---|---|
| Seat angle ... ... ... ... ... ... ... ... | 45° 30' |

Inlet:

| | |
|---|---|
| head diameter ... ... ... ... ... ... ... | 40.8 - 41.0 mm (1.606 - 1.614 in) |
| stem diameter ... ... ... ... ... ... ... | 7.972 - 7.987 mm (0.3136 - 0.3144 in) |

Exhaust:

| | |
|---|---|
| head diameter ... ... ... ... ... ... ... | 35.8 - 36.0 mm (1.409 - 1.417 in) |
| stem diameter ... ... ... ... ... ... ... | 7.965 - 7.980 mm (0.3136 - 0.3142 in) |
| Valve guide inside diameter ... ... ... ... ... ... | 8.000 - 8.022 mm (0.3149 - 0.3158 in) |

Valve guide/stem clearance:

| | |
|---|---|
| inlet ... ... ... ... ... ... ... ... | 0.013 - 0.050 mm (0.0005 - 0.0019 in) |
| exhaust ... ... ... ... ... ... ... ... | 0.020 - 0.057 mm (0.0008 - 0.0022 in) |

Valve guide outside diameter:

| | |
|---|---|
| inlet ... ... ... ... ... ... ... ... | 14.064 - 14.075 mm (0.5537 - 0.5541 in) |
| exhaust ... ... ... ... ... ... ... ... | 14.107 - 14.118 mm (0.55541 - 0.5558 in) |
| Valve guide housing inside diameter ... ... ... ... | 14.000 - 14.018 mm (0.5512 - 0.5519 in) |

Valve spring free length:

| | | |
|---|---|---|
| outer ... ... ... ... ... ... ... ... ... | 52.6 mm (2.07 in) |
| inner ... ... ... ... ... ... ... ... ... | 44.7 mm (1.779 in) |

## Rocker arms

| | |
|---|---|
| Bush inside diameter ... ... ... ... ... ... | 15.032 - 15.059 mm (0.5918 - 0.5929 in) |
| Spindle outside diameter ... ... ... ... ... | 14.982 - 14.994 mm (0.5899 - 0.5903 in) |
| Bush/spindle clearance ... ... ... ... ... ... | 0.038 - 0.079 mm (0.0015 - 0.003 in) |

## Camshaft

Journal diameter:

| | |
|---|---|
| flywheel side ... ... ... ... ... ... ... | 31.984 - 32.000 mm (1.2592 - 1.2598 in) |
| timing side ... ... ... ... ... ... ... | 46.984 - 47.000 mm (1.814 - 1.850 in) |

Housing diameter:

| | |
|---|---|
| flywheel side ... ... ... ... ... ... ... | 32.025 - 32.050 mm (1.2607 - 1.2623 in) |
| timing side ... ... ... ... ... ... ... | 47.025 - 47.050 mm (1.8511 - 1.8529 in) |

Clearance:

| | |
|---|---|
| flywheel side ... ... ... ... ... ... ... | 0.025 - 0.066 mm (0.001 - 0.0035 in) |
| timing side ... ... ... ... ... ... ... | 0.025 - 0.066 mm (0.001 - 0.0035 in) |

## Tappets

Guide bore diameter:

| | |
|---|---|
| standard ... ... ... ... ... ... ... ... | 22.021 - 22.000 mm (0.8669 - 0.8661 in) |
| 1st oversize ... ... ... ... ... ... ... ... | 22.071 - 22.050 mm (0.8689 - 0.8681 in) |
| 2nd oversize ... ... ... ... ... ... ... ... | 22.121 - 22.100 mm (0.8709 - 0.8700 in) |

Tappet diameter:

| | |
|---|---|
| standard ... ... ... ... ... ... ... ... | 21.996 - 21.978 mm (0.8695 - 0.8652 in) |
| 1st oversize ... ... ... ... ... ... ... ... | 22.046 - 22.028 mm (0.8679 - 0.8672 in) |
| 2nd oversize ... ... ... ... ... ... ... ... | 22.096 - 22.018 mm (0.8699 - 0.8668 in) |
| Clearance ... ... ... ... ... ... ... ... | 0.004 - 0.043 mm (0.0015 - 0.0016 in) |

## Crankshaft

Main bearing journal diameter:

| | |
|---|---|
| flywheel side ... ... ... ... ... ... ... | 53.970 - 53.931 mm (2.1248 - 2.1233 in) |
| timing side ... ... ... ... ... ... ... | 37.975 - 37.959 mm (1.4951 - 1.4944 in) |

Main bearing inside diameter:

| | |
|---|---|
| flywheel side ... ... ... ... ... ... ... | 54.000 - 54.019 mm (2.1260 - 2.1267 in) |
| timing side ... ... ... ... ... ... ... | 38.000 - 38.016 mm (1.4961 - 1.4967 in) |

Clearance:

| | |
|---|---|
| flywheel side ... ... ... ... ... ... ... | 0.030 - 0.068 mm (0.0011 - 0.0027 in) |
| timing side ... ... ... ... ... ... ... | 0.025 - 0.057 mm (0.0010 - 0.0022 in) |

**Note:** Main bearings are available in three oversizes of 0.20 mm (0.0079 in) increments

Big-end journal diameter:

| | |
|---|---|
| Class A ... ... ... ... ... ... ... ... | 44.008 - 44.014 mm (1.7325 - 1.7328 in) |
| Class B ... ... ... ... ..., ... ... ... | 44.014 - 44.020 mm (1.7328 - 1.7330 in) |

**Note:** On original assembly the big-end bearings are selectively matched with two classes of journal sizes. Class A marked white is matched with class B crankshaft, marked white. Class B bearings marked blue on the connecting rod are matched with class A crankshaft, marked blue. When the crankshaft requires re-grinding three bearing oversizes are available.

Big-end bearing:

| | |
|---|---|
| 1st oversize ... ... ... ... ... ... ... ... | 0.254 mm (0.010 in) |
| 2nd oversize ... ... ... ... ... ... ... ... | 0.508 mm (0.020in) |
| 3rd oversize ... ... ... ... ... ... ... ... | 0.762 mm (0.030 in) |

| | |
|---|---|
| Big-end bearing clearance ... ... ... ... ... | 0.030 - 0.054 mm (0.0011 - 0.0021 in) |
| Big-end axial float ... ... ... ... ... ... | 0.030 - 0.040 mm (0.0011 - 0.0015 in) |

| | |
|---|---|
| Small end bearing | |
| Small end bush diameter ... ... ... ... ... ... | 22.025 - 22.045 mm (0.8670 - 0.8767 in) |
| Gudgeon pin diameter ... ... ... ... ... ... | 22.000 - 22.004 mm (0.8661 - 0.8663 in) |
| Clearance ... ... ... ... ... ... ... ... | 0.021 - 0.045 mm (0.0008 - 0.0017 in) |

## Cylinder barrel diameter

| | |
|---|---|
| Class A ... ... ... ... ... ... ... ... | 83.000 - 83.006 mm (3.2670 - 3.2679 in) |
| Class B ... ... ... ... ... ... ... ... | 83.006 - 83.012 mm (3.2679 - 3.2681 in) |
| Class C ... ... ... ... ... ... ... ... | 83.012 - 83.018 mm (3.2681 - 3.2684 in) |

**Note:** Cylinders are selectively matched on original assembly with pistons of same class. Because the cylinder bore is chrome plated no subsequent rebore is possible

### Pistons and rings

Piston outside diameter

| | |
|---|---|
| Class A ... ... ... ... .... ... ... ... | 82.968 - 82.974 mm (3.2664 - 3.2668 in) |
| Class B ... ... ... ... ... ... ... ... | 82.974 - 82.980 mm (3.2668 - 3.2669 in) |
| Class C ... ... ... ... ... ... ... ... | 82.980 - 82.986 mm (3.2669 - 3.2671 in) |

Ring end gap:

| | |
|---|---|
| compression rings ... ... ... ... ... ... ... | 0.30 - 0.45 mm (0.0118 - 0.0180 in) |
| oil control ring ... ... ... ... ... ... ... | 0.25 - 0.40 mm (0.010 - 0.0150 in) |

Side clearance:

| | |
|---|---|
| compression rings ... ... ... ... ... ... ... | 0.030 - 0.062 mm (0.0011 - 0.0024 in) |
| oil control ring ... ... ... ... ... ... ... | 0.025 - 0.040 mm (0.0010 - 0.0015 in) |

## 850 Le Mans model

### Engine

| | |
|---|---|
| Type ... ... ... ... ... ... ... ... ... | Air cooled 90° V-twin four stroke |
| Bore ... ... ... ... ... ... ... ... ... | 83 mm (3.267 in) |
| Stroke ... ... ... ... ... ... ... ... | 78 mm (3.070 in) |
| Capacity ... ... ... ... ... ... ... ... | 844 cc (51.49 in) |
| Compression ratio ... ... ... ... ... ... | 10.2 : 1 |
| bhp (SAE) ... ... ... ... ... ... ... ... | 80 @ 7,300 rpm |
| Max rpm ... ... ... ... ... ... ... ... | — |
| Direction of rotation ... ... ... ... ... ... | Clockwise, viewed from front |

**Note:** All data, except for the following, is as given for 850T, T3 and California models

### Cylinder barrel diameter

| | |
|---|---|
| Class A ... ... ... ... ... ... ... ... ... | 83.000 - 83.009 mm (3.2677 - 3.2680 in) |
| Class B ... ... ... ... ... ... ... ... ... | 83.009 - 83.018 mm (3.2680 - 3.2684 in) |

**Note:** Cylinders are selectively matched to piston of same class on original assembly. On subsequent rebore two oversizes of piston are available

### Pistons and rings

Piston diameter:

| | |
|---|---|
| Class A ... ... ... ... ... ... ... ... | 82.936 - 82.945 mm (3.2651 - 3.2655 in) |
| Class B ... ... ... ... ... ... ... ... | 82.945 - 82.954 mm (3.2655 - 3.2660 in) |

Piston oversize:

| | |
|---|---|
| 1st oversize ... ... ... ... ... ... ... ... | 0.40 mm (0.0157 in) |
| 2nd oversize ... ... ... ... ... ... ... ... | 0.60 mm (0.023 in) |

## 750S and 750S3 models

### Engine

| | |
|---|---|
| Type ... ... ... ... ... ... ... ... ... | Air cooled 90° V-twin, four stroke |
| Bore ... ... ... ... ... ... ... ... ... | 82.5 mm (3.247 in) |
| Stroke ... ... ... ... ... ... ... ... ... | 70.0 mm (2.756 in) |
| Capacity ... ... ... ... ... ... ... ... | 748.4 cc (45.66 cu in) |
| Compression ratio ... ... ... ... ... ... ... | 9.8 : 1 |
| bhp (SAE) ... ... ... ... ... ... ... ... | 70 @ 7,000 rpm |
| Direction of rotation ... ... ... ... ... ... | Clockwise, viewed from front |

**Note:** All data, except for the following, is as for 850T and T3 models

### Valve timing

| | |
|---|---|
| Inlet opens ... ... ... ... ... ... ... ... | 40° BTDC |
| Inlet closes ... ... ... ... ... ... ... ... | 70° ABDC |
| Exhaust opens ... ... ... ... ... ... ... ... | 63° BBDC |
| Exhaust closes ... ... ... ... ... ... ... ... | 29° ATDC |

### Cylinder barrel diameter

| | |
|---|---|
| Class A ... ... ... ... ... ... ... ... ... | 82.500 - 82.506 mm (3.2480 - 3.2482 in) |
| Class B ... ... ... ... ... ... ... ... ... | 82.506 - 82.512 mm (3.2482 - 3.2484 in) |
| Class C ... ... ... ... ... ... ... ... ... | 82.512 - 82.516 mm (3.2484 - 3.2486 in) |

### Pistons and rings

Piston diameter

| | |
|---|---|
| Class A ... ... ... ... ... ... ... ... | 82.458 - 82.464 mm (3.2463 - 3.2465 in) |
| Class B ... ... ... ... ... ... ... ... | 82.464 - 82.470 mm (3.2465 - 3.2467 in) |
| Class C ... ... ... ... ... ... ... ... | 82.470 - 82.476 mm (3.2467 - 3.2469 in) |

## 1 General description

The engines fitted to all the Moto Guzzi models covered in this manual are of similar configuration and basic design and share many identical components. The engine is a 90° V-twin mounted in line with the frame, so that the angled cylinders are positioned in the direct air flow for efficient cooling. To aid this and maintain lightness, all major castings are in aluminium alloy. The crankshaft is a one-piece unit supported on plain bearings at the front and rear. Both bearings are one-piece components integral with their cast aluminium housings. The H section connecting rods have split big ends with detachable shell bearings, the connecting rods being supported on a single journal.

A single camshaft driven by chain from, and mounted above the crankshaft, operates the valves via steel cup-type followers and aluminium pushrods. The camshaft incorporates the tachometer drive gear and the contact breaker cam drive gear at the front and rear respectively.

On V-1000 models, the aluminium cylinder barrels are fitted with cast iron liners, which may be over bored. All other models have chromed-plated bores which extend considerably the expected bore life, but which must be renewed when wear becomes excessive. Overboring is not possible.

Lubrication is provided by a helical gear oil pump driven by the camshaft chain. Oil is picked up from the sump via a mesh filter and is supplied under pressure to all the working parts of the engine, after passing through a cartridge type filter (not 750s and 850T models). The oil is returned to the sump by gravity.

## 2 Operations with engine unit in the frame

1  It is only necessary to remove the engine from the frame if the crankshaft or main bearings require attention.
2  Most other work may be undertaken fairly easily with the engine still in the frame.
3  Since removing the engine, which is heavy, requires partial dismantling of the rear drive and suspension, it is advantageous to do all such work with the engine in situ. If, however a major overhaul is to be undertaken when a large number of components must be detached, removal of the engine/gearbox unit will aid accessibility and general ease of working.
4  The following components may be removed with the engine in place:

*Cylinder heads, barrels and pistons*
*Connecting rods and big-end bearings.*
*Camshaft*
*Alternator*
*Oil pump*

## 3 Method of engine/gearbox removal

Unlike most motorcycles, the engine fitted to the Moto Guzzi cannot be lifted from the frame in the normal manner. Due to the configuration of the cylinders and to the close proximity of the engine/gearbox assembly to the frame tubes, it is necessary to lift the frame up and off the engine/gearbox unit, the latter assembly remaining on the ground supported by the sub-frame and rear stand. This can be accomplished after detaching the sub-frame from the main frame and removing the rear suspension swinging arm. In addition, all components interconnecting the engine with the frame must be detached or disconnected in the normal way. The engine and gearbox can be removed only as a unit and not as individual components.

## 4 Removing the engine/gearbox unit from the frame

1  Place the machine on the centre stand so that it is supported securely on level gound. Position a suitable container of more than 3.5 litres (7.3/6.0 US/Imp pints) capacity below the engine and remove the drain plug from the rear wall of the sump. If work is to be carried out on the gearbox, this too should be drained. Place a suitable container of at least 0.8 litre (1.7 US pt/1.4 Imp pt) capacity beneath the gearbox oil drain plug; remove the plug and allow the oil to drain. In both cases, oil drainage will be accelerated if the engine has been allowed to reach full working temperature first; the oil will be thinner and so flow more readily.
2  On V-1000 models the torque converter hydraulic fluid must be drained from the system. Remove the left-hand frame side cover to gain access to the fluid tank. The cover is located by hooks at the upper edge and by a push fit projection at the base. Unscrew the lower banjo bolt from the fluid reservoir and allow the fluid to drain from the reservoir and pipe. Note the filter screen incorporated in the banjo bolt. Detach the hydraulic fluid return pipe from the underside of the bell housing. Drain also the fluid in the oil cooler by detaching the lower of the two pipes on the right-hand side of the gearbox. The pipe union is secured by a domed nut.
3  On all touring models the panniers and pannier crashbars must be removed to gain access during further dismantling. Removal of the handlebar screen is also recommended, to prevent accidental damage to this component during frame lifting operations. The removal of these ancillaries is quite straightforward, all being secured by brackets and bolts.
4  Raise the dualseat so that it is held in the upper-most position by the support rod provided, and lift out the tool tray. Remove the frame side covers, each of which is retained by hooks at the upper edge and by a push fit projection passing into a frame mounted grommet at the base. Unclip the battery retaining strap and detach both main leads from the battery terminals. The battery is very heavy and is a tight fit between the frame members and adjacent cycle parts. Due to this removal can be difficult. The battery must be tilted to effect removal. **DO NOT** allow battery acid to spill onto the frame or other components.
5  Detach the strap securing the rear of the petrol tank to the frame. Ensure that both petrol taps are closed and then disconnect the petrol feed pipes at the tap unions. The pipes are retained either by screw clips or by spring clips. On models fitted with the electro-valve tap, disconnect the wire running to the solenoid body, integral with the tap. V-1000 models utilise a fuel level sensor which is screwed into the underside of the petrol tank. The sensor wire must be disconnected; it is a push fit.

Drainage of the petrol is not strictly necessary, although it will reduce overall tank weight and so facilitate removal. The tank may be lifted rearwards and away from the machine to clear the two front mounting bolts which pass into either side of the steering head lug. Note the mounting rubbers.
6  Loosen the clamps which hold the left-hand exhaust pipe and silencer at the connections with the H section balance pipe. Pull the silencer from position after removing the single silencer support bolt. Each exhaust pipe is retained at the exhaust port by a finned flange, secured on the cylinder head by two studs and nuts. The flange pulls down on to a split ring collar, which holds the pipe in place. Remove the two nuts from the left-hand flange and then pull the exhaust pipe from position on the cylinder head and balance pipe. The right-hand exhaust system components, together with the balance pipe, may be detached as a unit after removing the flange and silencer mounting bolt. Le Mans exhaust pipes are one-piece, having a non detachable balance pipe. Remove the two pipe. Remove the two pipes as a unit.
7  Remove the two screws which hold each carburettor top in place. Pull out each throttle valve assembly and carburettor top by grasping the control cable. Displace the return springs, disconnect the throttle valves from their respective cables and slide the carburettor caps from position. Refit the throttle valves and carburettor tops to their respective carburettors to prevent accidental interchanging of the matched parts. Temporarily, tie

the throttle cables to some convenient part of the frame to prevent snagging during further dismantling operations. Where cable controlled choke assemblies are utilised, remove the screw holding each assembly in position and pull the units from the carburettors, complete with cables. Detach the control lever assembly from the left-hand rocker box, where it is retained by a single socket screw.

8  Disconnect the interconnected petrol feed pipes from each carburettor by removing the single screw passing through the unions. Removal of the unions gives access to the filter screens. These should be removed now, to prevent loss. The carburettors are then free to be detached. On all models using an air filter, carburettor removal is accomplished most easily by detaching each inlet stub at the cylinder head and pulling the stubs complete with carburettors from the rubber intake duct. If required, the carburettors may be detached from the stubs by loosening the clamps. Each stub is retained by three socket screws passing through a flange. Note that the lower screw also retains the HT lead guide clip.

9  Where fitted, the air filter duct is retained on the breather box by a strap arrangement consisting of two steel strips and two long springs. Unhook the springs and pull the duct from position. Disconnect the four pipes leading to the breather box. All four pipes are a push fit on the individual unions. The breather box and air filter element may be removed after unscrewing the single nut at the forward end of the air filter box. Removal of the air filter box itself is not possible until the frame has been lifted from the engine. This is due to the limited clearance between the engine and adjacent frame tubes.

10 Remove the alternator cover from the extreme front of the engine. The cover is retained by four screws. Disconnect the push fit leads from the alternator stator and free the wiring grommet from the casing. Temporarily refit the cover, to protect the alternator. Disconnect the following wires at the components listed:

*Oil pressure switch*
*Neutral indicator switch (not V-1000 model)*
*Side stand parking switch (V-1000 models only)*
*Starter motor solenoid*
*Contact breaker to coil leads*

11 Apply the handlebar clutch lever and, using a wooden lever between the frame tube and clutch operating arm, (at the gearbox) detach the clutch control cable. A safety switch is interconnected with the clutch cable close to the abutment on the gearbox. The switch allows engine starting only when the clutch is disengaged. Prise the rubber boot away from the switch and detach the two push fit leads. Free the clutch cable from the gearbox abutment.

12 Disconnect the speedometer drive cable from the gearbox

and the tachometer drive cable (where utilised) from the front of the timing cover. Both cables are retained by knurled rings. Note that the small crimped olive fitted to each cable end below the knurled ring is often loose, and so easily lost if care is not exercised.

13 On touring models the footboards should be removed. Each is retained by two pivot bolts, which have a nut and locknut.

14 Remove the clevis pin and split pin from both the gearchange link rod and rear brake operating rod (disc brake only). On touring models the shorter link rod should be disconnected. Removal of the operating levers is not strictly necessary but may improve working room. On machines fitted with a drum rear brake, the brake pedal should be removed from the splined pivot shaft. The pedal is secured by a pinch bolt.

15 Pull the push fit breather tube from the union immediately above the starter motor (not V-1000 models). Remove the nut and bolt which secures the lower flange of the starter motor to the engine casing. Support the weight of the starter motor, unscrew the upper bolt and lift the starter motor to the rear and clear of the machine.

16 Remove the seven screws which secure the battery carrier plate to the gearbox and frame, and lift out the plate. Note that the rear left-hand bolt secures the battery earth strap.

17 The rear wheel and the swinging arm must now be removed to allow disconnection of the final drive shaft from the gearbox output shaft. Refer to Chapter 5, Section 8 for the relevant details.

18 Place a number of wooden blocks below the front of the engine sump so that the weight of the engine is just taken. Remove the two socket screws and nuts which clamp together the lower ends of the front frame down tubes and the forward end of the two subframe tubes. On touring models these screws also secure the lower ends of the crashbars. The crashbars can be removed after unscrewing the upper securing bolts. Remove the nut from the front engine mounting bolt and carefully drive the bolt from position. Care should be taken not to damage the threads. On all but touring models the side stand is also secured by this bolt.

19 The sub frame - attached to the engine - is now free from the main frame, which can be lifted up from the rear and wheeled away on the front wheel. It is recommended that two assistants be at hand during this operation, one of which can steer the frame via the handlebars, the other to steady the engine/gearbox unit.

20 Once the frame has been lifted away, the sub frame may be detached by removal of the single bolt which passes through the two frame tube lugs and the gearbox mounting lug. Take care when removing the sub frame that the rear stand does not retract suddenly and catch the unwary individual across the knuckles.

4.1 Remove filler plug and drain plug and allow oil to drain

4.4 Detach strap and terminal and lift battery away

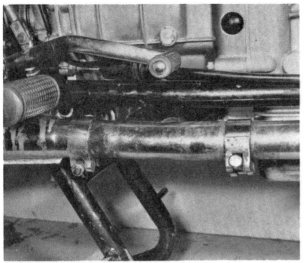
4.6a Loosen all exhaust clamps and ...

4.6b ... remove the two nuts from each flange

4.6c Pull the flange from the studs to release pipe

4.7a Unscrew carburettor tops and pull out slides

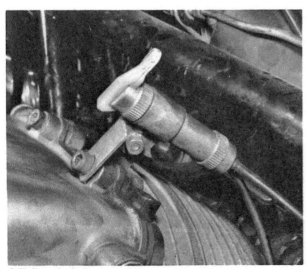
4.7b Detach choke control from rocker cover

4.8a Disconnect carburettor feed pipes

4.8b Remove carburettors complete with inlet stubs

4.9a Pull hoses from breather box unions

4.9b Tape air filter box to frame during removal

4.10a Note alternator lead positions and disconnect

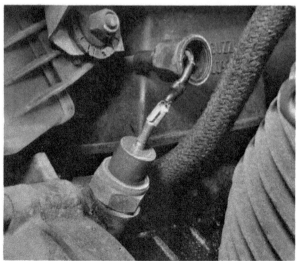

4.10b Disconnect oil pressure switch lead and ...

4.10c ... neutral indicator leads

4.10d Remove also starter motor leads

4.10e Pull contact breaker leads from coils ·

4.11 Prise boot off clutch cable switch to detach wires

4.12a Unscrew tachometer cable followed by ...

4.12b ... speedometer cable at gearbox

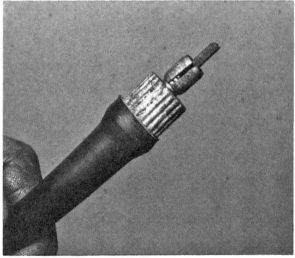

4.12c Do not loose the small olive

4.14. Remove clevis pin to separate gearchange link rod

4.15 Detach starter motor, held by two bolts

4.16a When removing battery plate bolts note ...

4.16b ... the bolt which holds the battery earth lead

4.18 Crashbars held at top by two bolts

4.19 Lift frame from engine with great care

## 5  Dismantling the engine/gearbox: general

1   Before commencing work on the engine, the external surfaces must be cleaned thoroughly. A motor cycle engine has very little protection from road dirt which will sooner or later find its way into the dismantled engine if this simple precaution is not observed.

2   One of the proprietary engine cleaning compounds such as Gunk or Jizer can be used to good effect, especially if the compound is allowed to penetrate the film of oil and grease before it is washed away. When washing down, make sure that water cannot enter the carburettors or the electrical system, particularly if these parts are now more exposed.

3   Never use force to remove any stubborn part, unless mention is made of this requirement in the text. There is invariably good reason why a part is difficult to remove, often because the dismantling operation has been tackled in the wrong sequence.

## 6  Separating the engine from the gearbox

1   The gearbox, which is retained on the rear of the engine by radially disposed nuts and studs, can be separated from the engine as a complete unit after removing these nuts.

2   On V-1000 models, it is ESSENTIAL that the engine/gearbox assembly is placed on the workbench, front end downwards with the crankshaft approximately vertical, before separation is accomplished. If this precaution is not observed, the fluid contained within the torque converter will spill out as the gearbox input shaft boss leaves the centre of the converter casing.

3   After separation, the torque converter fluid must be drained. The simplest method is to lift up and tip the engine unit, sufficiently to allow all the fluid to drain away. Alternatively, a small syphon pump can be used. A bicycle pump with the leather plunger cap inverted also makes a useful tool. After removing the fluid, place a cover over the torque converter orifice, to prevent the ingress of foreign matter.

## 7  Dismantling the engine: removing the cylinder heads, cylinder barrels and pistons

1   Each cylinder barrel/head assembly should be dismantled individually and the components stored separately, to prevent inadvertent interchanging of the matched parts. In addition, parts should be marked clearly so that they may be refitted in their original positions.

2   Commence dismantling by disconnecting the rocker oil feed pipes at the crankcase and at each cylinder head. The pipes are interconnected by a shared banjo union at the crankcase and by separate banjo unions at the cylinder heads.

3   Unscrew the eight socket screws which retain the rocker cover in place. Lift off the cover and remove the gasket. Remove both spark plugs and rotate the engine so that both valves on the cylinder in question are closed, ie, the piston is at TDC on the compression stroke. Remove the screw locating the exhaust rocker arm spindle in the cast bracket. Push out the spindle and remove the rocker arm, bronze washer and coiled spring washer. Lift out the pushrod. Repeat the procedure for the inlet valve components.

4   The cylinder head is retained by six nuts, one of which is of the socket type. Access to this nut can be gained only after removal of the socket cap screwed into the cylinder head. Loosen and remove the nuts and washers and lift off the rocker carrier bracket. Each of the four rocker carrier securing studs is fitted with a small 'O' ring. Using a small screwdriver ease the 'O' rings from position. If difficulty is encountered in removing the 'O' rings they may be cut through, as they will have to be discarded.

The cylinder head is now free to be removed. If necessary, a rawhide mallet may be used to break the seal between the cylinder head and the cylinder barrel and gaskets. Take care not to damage the aluminium fins.

5   Separate the cylinder barrel from the crankcase mouth, if necessary using a rawhide mallet to break the seal. Again take care not to damage the fins. **Do not use levers to displace the cylinder barrel.** Slide the cylinder barrel up along the holding down studs until the piston skirt is visible, but the piston rings are still in the cylinder bore. At this stage it is worth padding the crankcase mouth with clean rag to prevent pieces of broken piston ring or other foreign matter from falling into the crankcase. Support the piston and pull the cylinder barrel clear of the studs and piston. Remove the cylinder base gasket and note the two 'O' rings one of which is fitted to the upper stud and the other to the lower stud.

6   Before removing the piston a mark should be scribed on the inside of the skirt to identify to which cylinder the piston belongs. An arrow on the piston crown accompanied by the letters SCA indicates the front of the piston. Remove the gudgeon pin circlips, using a pair of snipe nose pliers. The gudgeon pin may be pushed out using a special gudgeon pin extractor tool, or drifted out using a suitable soft brass drift. If the latter technique is adopted, the connecting rod must be supported securely against the side thrust imposed. Failure to ensure this, may lead to a bent connecting rod or a damaged big-end bearing.

7   If the gudgeon pin is very tight, the piston should be heated, using a rag soaked in boiling water, or a flat iron applied to the piston crown. This will temporarily expand the aluminium piston.

8   Repeat the dismantling procedure on the second cylinder barrel/head assembly.

## 8  Dismantling the engine: removing the alternator

1   Remove the alternator cover which was replaced temporarily during earlier dismantling operations. Very carefully lift the two alternator brushes half out of their holders, securing them in this position by off-setting the brush springs. Remove the three long retaining screws and pull the alternator stator from position.

2   The alternator rotor is a tight fit on the tapered crankshaft end and will require drawing from position. **Under no circumstances should levers be employed in an attempt to remove the rotor** as damage will almost certainly result. The rotor is retained by a central socket screw which may also be used as the extractor. Remove the bolt and insert a piece of 2¼ x 3/16 in rod into the drilling in the crankshaft end. Refit the bolt and tighten it down very slowly until the rotor leaves the shaft. If the rotor is reluctant to move, **DO NOT continue tightening the bolt.** A smart tap on the bolt end with a hammer should release the rotor.

3   Store the alternator components in a safe place, to prevent damage.

6.1 Gearbox secured to engine by studs and bolts

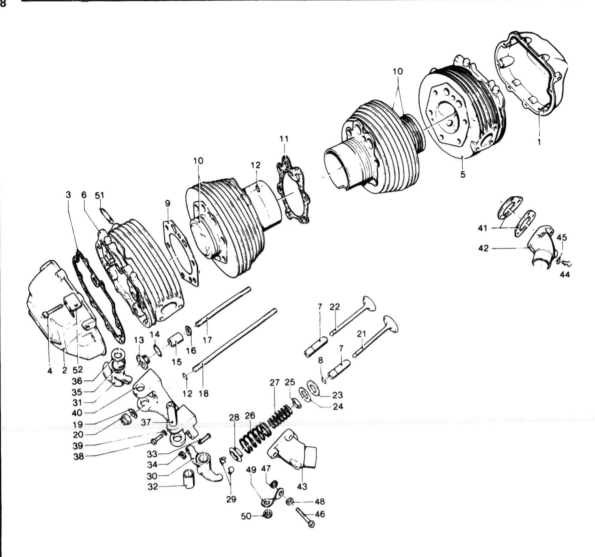

**Fig. 1.1. Cylinder head and cylinder barrel - components**

| | | | |
|---|---|---|---|
| 1 | RH rocker cover | 27 | Inner spring - 4 off |
| 2 | LH rocker cover | 28 | Spring collar - 4 off |
| 3 | Gasket - 2 off | 29 | Collet set - 4 off |
| 4 | Socket screw - 16 off | 30 | RH rocker arm - 2 off |
| 5 | RH cylinder head | 31 | LH rocker arm - 2 off |
| 6 | LH cylinder head | 32 | Rocker bush - 4 off |
| 7 | Valve guide - 4 off | 33 | Tappet adjuster - 4 off |
| 8 | Circlip - 4 off | 34 | Locknut - 4 off |
| 9 | Gasket - 2 off | 35 | Endfloat spring - 4 off |
| 10 | Cylinder barrel and | 36 | Bronze washer - 4 off |
| | piston assembly - 2 off | 37 | Rocker shaft - 4 off |
| 11 | Base gasket - 2 off | 38 | Locating screw - 4 off |
| 12 | 'O' ring - 10 off | 39 | Spring washer - 4 off |
| 13 | Blanking plug - 25 off | 40 | Rocker bracket - 2 off |
| 14 | 'O' ring - 25 off | 41 | Heat sink - 2 off |
| 15 | Sleeve socket nut - 2 off | 42 | RH inlet stub |
| 16 | Wave washer - 2 off | 43 | LH inlet stub |
| 17 | Short stud - 4 off | 44 | Socket screw - 2 off |
| 18 | Long stud - 8 off | 45 | Plain washer - 6 off |
| 19 | Washer - 10 off | 46 | Socket screw - 6 off |
| 20 | Cylinder head nut - 10 off | 47 | Insulating washer - 6 off |
| 21 | Inlet valve - 2 off | 48 | Plain washer - 6 off |
| 22 | Exhaust valve - 2 off | 49 | Cable guide - 2 off |
| 23 | Washer - 4 off | 50 | Grommet - 2 off |
| 24 | Washer - 10 off | 51 | Stud - 4 off |
| 25 | Spring seat - 4 off | 52 | Choke lever bracket |
| 26 | Outer spring - 4 off | | |

7.2a Disconnect the rocker feed pipe at each cylinder head ...

7.2b ... and at the central union on the crankcase

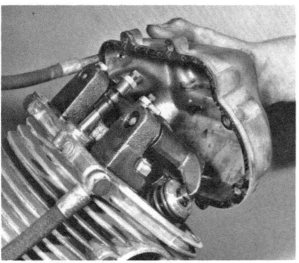

7.3a Lift off the rocker cover

7.3b Unscrew the locating bolts and ....

7.3c ... push out the rocker spindles

7.3d Lift pushrods from position

7.3e Access to the four inner nuts is now possible

7.4a Remove the internal socket cap to ...

7.4b ... enable upper sleeve bolt to be unscrewed

7.4c Do not forget lower nut near spark plug

7.4d Prise the 'O' rings from the long studs

7.4e Slide cylinder head off the studs

7.5 Displace the cylinder barrel and lift off

7.6 Prise out the circlips and ...

7.7 Push out gudgeon pin to free the piston

8.1a Remove alternator stator held by three screws

8.2a After removing centre screw insert short rod

8.2b Rotate centre screws to push rotor off shaft taper

## 9  Dismantling the engine: removing the timing cover, timing chain and sprockets

1   Arrange the engine so that it is resting on the gearbox mounting studs. Loosen and remove the timing cover screws and lift the cover from position. The use of a rawhide mallet may be required to loosen the cover from the gasket.
2   Unscrew the camshaft sprocket nut, oil pump sprocket nut and crankshaft end nut. The latter component is of the ring type and is secured by a special tab washer. Bend down the ears of the washer before attempting to loosen the nut. This nut requires a peg type spanner for removal and refitting. A suitable tool is fabricated easily from a thick walled tube, one end of which is relieved with a file to form four short pegs.
3   To prevent rotation when loosening these three nuts, pass a close fitting steel rod through the small end eye of one connecting rod. Allow the rod to bear on two wooden blocks placed across the crankcase mouth.
4   Before the cam chain is detached, rotate the engine until the right-hand piston is at TDC on the compression stroke. Using a centre punch or scribe, mark the relative positions of the distributor body and crankcase, the body and contact breaker mounting plate and the contact breaker cam and distributor body. This will simplify ignition timing and distributor replacement.
5   The three chain sprockets must be lifted off their shafts simultaneously, together with the camshaft drive chain. The camshaft sprocket is a light push on the camshaft end and will lift off with ease. The remaining sprockets however may require careful easing from position, using short levers. Care should, of course, be taken to ensure no damage is done to any of the components against which the levers bear.
6   Remove the push fit drive pin from the camshaft end boss and prise the Woodruff key from the keyway in each of the two other shafts. The keys are easily lost so store them in a safe place.
7   Unscrew the two bolts which secure the chain tensioner arm. Each bolt is secured by a locking plate and is fitted with a distance piece between the arm and the wall of the casing.

## 10  Dismantling the engine: removing the distributor, camshaft and engine oil pump

1   Loosen the two bolts which retain the distributor clamping bracket. Remove the front bolt and swing the bracket outwards to clear the distributor boss. The distributor should be marked

before removal as described in paragraph 4 of the preceding Section, to aid correct positioning when refitting. Lift the complete distributor from place.
2   Lift out the cam followers and mark each individual unit so that they may be replaced in their original positions. Detach the camshaft bearing end plate which is retained by three screws. Grasp the camshaft end and withdraw it, together with the end plate.
3   Loosen and remove the oil pump retaining socket screws and lift the oil pump from place. The oil pump is located accurately on two dowel pins and may be held securely in place. **DO NOT use levers to displace the pump.** Temporarily replace the oil pump sprocket and nut and use the sprocket as a means of purchase.

## 11  Dismantling the engine: removing the clutch and flywheel (except V-1000 Convert model)

1   The clutch components are retained by eight bolts passing through the periphery of the starter ring gear into the flywheel face. Unscrew the eight bolts and lift off the starter ring gear. Remove the following components consecutively; outer friction plate, intermediate plain plate, inner friction plate, clutch thrust piece, spring back plate and eight clutch springs. Because the complete assembly is under pressure from the clutch springs, the eight bolts must be unscrewed evenly, about one turn at a time, so that the spring pressure is released in a controlled and even manner.
2   Before removing the flywheel, note the white paint mark on the crankshaft end boss which aligns with the TDC mark on the flywheel. This mark was made on original assembly to ensure ease of correct flywheel positioning with regard to crankshaft and timing mark relationship.
3   Remove the flywheel retaining bolts after bending down the ears of the locking plates which secure the bolts in pairs.

## 12  Dismantling the engine: removing the torque converter and flywheel (V-1000 Convert model only)

1   The torque converter is secured by four bolts, of which each is secured by a tab washer. After removal of the bolts the complete converter may be lifted out, followed by the starter ring gear.
2   Flywheel removal follows the technique detailed for all other models.

9.1 Remove the timing cover and gasket

9.2a Bend down the multi-tab locking washer and ...

9.2b .. using a suitable tool remove the crankshaft nut

9.2c Remove the oil pump nut and ...

9.2d ... the camshaft end nut

9.5 Lift off the three sprockets and chain as a unit

9.7 Remove the chain tensioner, noting the spacers

10.2a Lift out each cam follower

10.2b Detach the camshaft end plate and ...

10.2c ... withdraw the camshaft

10.3 The oil pump is retained by four socket screws

11.1a Lift out clutch thrust piece

11.1b Unscrew the clutch bolts and remove the plates

11.2 Lift out the spring plate and displace the springs

11.3 Bend down tab washers to remove flywheel bolts

12.1 Torque converter is held by four secured bolts - V-1000

## 13 Dismantling the engine: removing the connecting rods and the crankshaft.

1   Invert the engine so that it is resting in an upright position on the cylinder head studs. Remove the sump bolts and lift the sump away, complete with the oil filter system.

2   Before removing the connecting rods, mark each rod and big-end cap set so that they are refitted as a matched pair and in the same position on the big-end journal. Unscrew the big-end nuts and separate the caps from the rods. Remove the connecting rods and temporarily refit the big-end caps. Do not allow the bearing shells to be interchanged.

3   Reposition the engine so that the front is facing downwards and is supported on blocks, with the crankshaft end clear of the workbench. Loosen and remove the bolts from the rear main bearing housing after bending down the tab washer ears. The bearing housing is now free to be removed. If required, the housing may be drifted out by applying a soft nosed mallet to the front end of the crankshaft.

4   Removal of the crankshaft is straightforward. Lift the complete assembly out through the rear of the crankcase, turning it as necessary so that the webs clear the edge of the casing. Unless the front main bearing requires renewal, removal of the bearing housing is not required. The housing is retained by six bolts, secured in pairs by locking plates.

13.2 Mark caps and rods before removing

13.3 Detach the rear main bearing housing and ...

13.4a ... manoeuvre the crankshaft from position

13.4b Front main bearing housing - general view

## 14 Examination and renovation: general

1   Before examining the parts of the dismantled engine for
wear, it is essential that they should be cleaned thoroughly. Use
a petrol/paraffin mix to remove all traces of old oil and sludge
that may have accumulated within the engine and a cleansing
agent such as Gunk or Jizer for the external surfaces. Special
care should be taken when using these latter compounds, which
require a water wash after they have had time to penetrate the
film of grease and oil. Water must not be allowed to enter any of
the internal oilways or parts of the electrical system.
2   Examine the crankcase castings for cracks or other signs of
damage. If a crack is discovered, it will require specialist repair,
or replacement.
3   Examine carefully each part to determine the extent of wear,
if necessary checking with the tolerance figures listed in the
Specifications Section of this Chapter. The following Sections of
this Chapter describe how to examine the various engine components for wear and how to decide whether renewal is necessary
4   Always use a clean, lint-free rag for cleaning and drying the
various components prior to reassembly, otherwise there is risk
of small particles obstructing the internal oilways.

## 15 Crankshaft, crankshaft bearings and connecting rods: examination and renovation

1   Check the big-ends for wear, whilst assembled on the crankshaft, by pushing and pulling on the connecting rods. There
should be no discernible play.
2   When dismantled, check each big-end width with a micrometer, and check with the dimension given in the Specifications.
If the big-end shells are badly scored, they will have to be
renewed. Endeavour to find out why they were scored - check
oilways etc.
3   Check that the small end bushes are secure, and have no more
than the allowable wear. If they need to be replaced, they must be
pressed out. The new bush may be used to press the old bush from
position. See the accompanying diagram. To restore the correct
clearance between the bush and gudgeon pin, the bushes must be
reamed out after fitting.
4   Inspect and measure the internal diameter of both main bearings and the diameters of the crankshaft journals. Renew the
bearings, complete with housings, if the clearance exceeds that
given in the Specifications. Ideally, if crankshaft or big-end wear
has occurred, the bearing journals should be re-ground undersize
and suitable bearings fitted. Some crankshafts are nitrided (a

special hardening process) however, and if no journal ovality,
scoring or tapering is found, regrinding may not be required.
Consult a Moto Guzzi specialist for advice on this matter. Oversizes for both main and big-end bearings are available, in three
categories.
5   It is essential that the main and big-end journals are re-
radiused whenever the crankshaft undergoes regrinding. If this
operation is overlooked, the stresses imposed may cause crankshaft failure. The correct radii are as followes:

| | |
|---|---|
| *Big-end bearing* | *2.0 - 2.5 mm (0.078 - 0.090 in)* |
| *Rear main bearing* | *3.0 mm (0.118 in)* |
| *Front main bearing* | *1.5 - 1.8 mm (0.058 - 0.070 in)* |

6   Check that the oilways in the crankshaft are quite clear. A
high pressure air hose is suitable for removing all normal deposits.
A centrifugal sludge trap is incorporated in the crankshaft, the
chamber of which is closed by a socket blanking-plug in the
forward crankshaft web. If removed, the plug must be refitted,
using a locking fluid applied to the thoroughly cleaned threads.
It is essential that the plug is quite tight as oil failure and other
damage will occur if it unscrews.

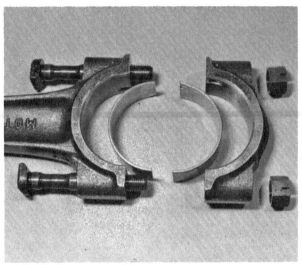

15.2 Connecting rod components - general view

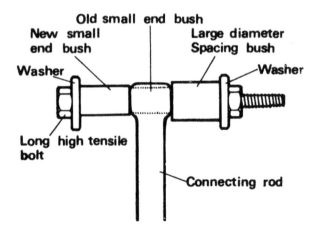

Fig. 1.2. Removal of small end bush

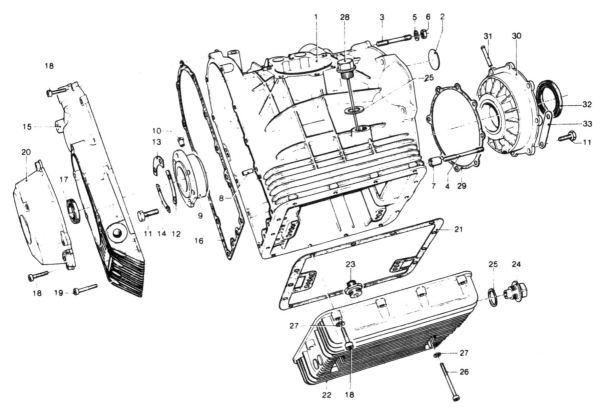

**Fig. 1.3. Crankcase - component parts**

| | | | |
|---|---|---|---|
| 1 | Crankcase | 18 | Socket screw - 26 off |
| 2 | Core plug | 19 | Socket screw - 6 off |
| 3 | Short stud - 4 off | 20 | Alternator cover |
| 4 | Long stud - 2 off | 21 | Gasket |
| 5 | Plain washer - 5 off | 22 | Jump |
| 6 | Nut - 6 off | 23 | Filter union - not |
| 7 | Hollow dowel - 2 off | | 750S or 850T |
| 8 | Hollow dowel - 2 off | 24 | Drain plug |
| 9 | Main bearing/housing | 25 | Washer |
| 10 | Oilway insert | 26 | Socket screw - 4 off |
| 11 | Bolt - 14 off | 27 | Plain washer - 18 off |
| 12 | Lock plate | 28 | Oil filler cap/dipstick |
| 13 | Lock plate | 29 | Gasket |
| 14 | Lock plate | 30 | Main bearing/housing |
| 15 | Timing cover | 31 | Oil pipe |
| 16 | Gasket | 32 | Oil seal |
| 17 | Oil seal | 33 | Lock plate |

**16 Camshaft and pushrods : examination and renovation**

1   The camshaft is unlikely to show signs of wear unless a high mileage has been covered or there has been a breakdown in the lubrication system. Wear will be most obvious on the flanks of the cams and at the peak, where flattening-off may occur. Scuffing or in an extreme case, discoloration, is usually/ indicative of lubrication breakdown.

2   If there is any doubt about the condition of the camshaft, it is advisable to renew it whilst the engine is completely dis-mantled. Comparison with a new camshaft is often the best means of checking visually the extent of wear.

3   Check the cam followers for wear or damage. Again it is extremely unlikely that any has occurred. Slight scoring may be removed using an oil stone, provided that the surface remains absolutely flat and square. The cam followers are designed to rotate slowly in order that wear is distributed evenly over the

surface of the working face. Lack of rotation will be evident by a marked depression where the cam lobe has rubbed consistently. Check the fit of the cam followers in their respective slide-ways in the crankcase. Excessive clearance may be restored by fitting new followers, of which two oversizes are available. In extreme cases the slide-ways in the crankcase may have to be reamed to restore the surfaces.

4   Check the pushrods for straightness by rolling them on a flat surface. Replace any that are bent, since it is impractical to straighten them with accuracy. Check that the hardened end pieces are not loose, or the internal bearing surfaces worn, chipped or broken.

5   Check the clearance between the camshaft journals and the bearing surfaces. The camshaft runs directly in the aluminium crankcase. Wear is usually negligible, a matter that is fortuitous, because wear in the bearings will require renewal of the complete crankcase unit.

## 17 Timing chain, sprockets and tensioner: examination and renovation

1   It is unlikely that either the timing chain or sprockets will require renewal, unless a chain breakage has damaged the teeth. Both the chain and sprockets are designed for long life.
2   Check for uneven wear of the chain when still mounted on the crankcase, by removing the tensioner and turning the crankshaft a quarter turn at a time. Measure the play in the chain at each turn. If in doubt about chain condition, it should be renewed, since breakage may damage the sprockets or the crankcase - apart from immobilising the machine.
3   The timing chain is endless, and should be examined carefully when removed for broken rollers or cracked side plates or rivets damaged when extracting the sprocket.
4   Inspect the teeth of the sprockets for chipping or hooking.
5   Check that the rubber slipper surface of the chain tensioner is not damaged. It may be grooved but as long as the rollers do not make contact it is still serviceable.

## 18 Crankcase and timing chain cover oil seals: examination and renovation

1   The presence of oil in the clutch housing, or alternator housing may indicate failure of the crankshaft oil seals. The damaged seal may be drifted or prised carefully from position.
2   The front oil seal is fitted in the timing case and the rear seal in the rear main bearing housing. When refitting the seals, the spring side must face towards the engine.
3   It is recommended that both seals be renewed when the engine is dismantled as subsequent failure will require considerable dismantling for renewal.

## 19 Cylinder barrels: examination and renovation

1   The usual indications of badly worn cylinder bores and pistons are excessive oil consumption and piston slap, a metallic rattle which occurs when there is little or no load on the engine. If the top of the cylinder barrel is examined carefully, it will be observed that there is a ridge on the thrust side of each cylinder bore which marks the limit of travel of the uppermost piston ring. The depth of this ridge will vary according to the amount of wear that has taken place and can therefore be used as a guide to bore wear.
2   On all models the pistons are selectively matched on initial assembly to fit the cylinder barrels, and are stamped A, B or C on the the piston crown. The cylinder barrels are marked likewise. When taking cylinder barrel or piston measurements refer to the classification in the relevant Specifications at the beginning of the Chapter.
3   With the exception of V-1000 and Le Mans models, all models are fitted with chrome-plated cylinder bores. This type of bore has very good wearing properties, consequently a long life may be expected. Because the chrome layer is very thin, no overboring is possible. If wear becomes excessive or damage occurs to the bore, the complete cylinder barrel must be renewed. V-1000 and Le Mans models are fitted with an aluminium alloy cylinder barrel, utilising a traditional steel sleeve bore. Two sizes of oversize piston are available for fitting after reboring in the normal manner.
4   Measure the bore diameter using an internal micrometer. Three measurements should be taken, at the top, middle and bottom of each bore. A further three measurements should then be taken at 90° to the first positions. On models fitted with chrome bores, wear should not exceed 0.1 mm (0.004 in) over the largest bore diameter given for that classification of cylinder barrel in the Specifications. If wear exceeds this figure, the cylinder barrels must be renewed. Wear on steel cylinder bores should not exceed the greatest diameter given for that classification of piston.

5   Check the surfaces of each cylinder bore to ensure there are no score marks or other signs of damage that may have resulted from an earlier engine seizure or displacement of one of the circlips. Even if the bore wear is not sufficient to necessitate renewal or a rebore, a deep indentation will override this decision in view of the compression leak that will occur.

## 20 Pistons and piston rings: examination and renovation

1   If a rebore is necessary, the pistons and rings can be discarded because they must be replaced by their oversize counterparts.
2   Remove all traces of carbon from the piston crown, using a soft scraper to ensure the surface is not marked. Finish off by polishing the crown with metal polish, so that the carbon will not adhere so readily. **Never** use emery cloth.
3   Piston wear usually occurs at the base of the skirt and takes the form of vertical streaks or score marks on the thrust side. If a previous engine seizure has occurred, the score marks will be very obvious. Pistons which have been subjected to heavy wear or seizure should be rejected and new ones obtained.
4   Remove the piston rings carefully, by expanding them sufficiently to pass over the piston. The rings are very brittle, and must not be handled roughly. Note which groove each ring came out of, and which way up on each piston.
5   Clean the ring grooves of any burnt deposits. A piece of old broken ring is useful for this, if used carefully.
6   The piston ring grooves may become enlarged in use, permitting the rings to have greater side float. It is unusual for this type of wear to occur on its own, but if the side float appears excessive, new pistons of the correct size should be fitted.
7   Piston ring wear can be checked by inserting the rings, one at a time, in the cylinder bore and pushing them down about 1½ inches with the base of the piston so that they rest squarely in the bore. Using a feeler gauge check the end gap against the Specifications. If the gap exceeds the figure given, the rings should be renewed. On chrome bore cylinder barrels, new rings should **not** be fitted to old bores as the rings will not bed down on the highly polished surface.

## 21 Valves, valve springs, and valve guides: examination and renovation

1   Use a valve spring compressor to release each of the valves in turn. Keep the valves, valve springs and collets etc together in sets so that they are eventually replaced in their original location. Fitted below each spring lower seat are a number of shims and washers. Note the number of components used for each valve together with their relative positions.
2   After cleaning all four valves to remove carbon and burnt oil, examine the heads for signs of pitting or burning. Examine the valve seats in the cylinder head. The exhaust valves and their seats will require the most attention because they are the hotter running. If the pitting is slight, the marks can be removed by grinding the seats and valve heads together, using fine valve grinding compound.
3   Valve grinding is a simple, if somewhat laborious task. Smear a trace of fine valve grinding compound (carborundum paste) on the seat face and apply a suction grinding tool to the head of the valve. Oil the stem of the valve and insert it in the guide until it seats in the grinding compound. Using a semi-rotary motion, grind-in the valve head to its seat. Lift the valve occasionally to distribute the grinding compound more evenly. Repeat this application until an unbroken ring of light grey matt finish is obtained on both valve and seat. This denotes the grinding operation is now complete. Before passing to the next valve, make sure that all traces of the valve grinding compound have been removed from both the valve and its seat and that none has entered the valve guide. If this precaution is not observed, rapid wear will take place due to the highly abrasive nature of the carborundum base.

4   When deep pits are encountered, it will be necessary to use a valve refacing machine and a valve seat cutter, set to an angle of 45°. Never resort to excessive grinding because this will only pocket the valve in the head and lead to reduced engine efficiency. If there is any doubt about the condition of a valve, fit a new one.

5   Examine the condition of the valve collets and the grooves in the valve stem in which they seat. If there is any sign of damage, new parts should be fitted. Check that the valve spring collar is not cracked. If the collets work loose or the collar splits whilst the engine is running, a valve could drop in and cause extensive damage.

6   Measure the valve stems for wear, comparing them with the unworn portion that does not extend into the valve guide. Check also the valve guides for excessive play. Check that the end of the stem is not indented from contact with the rocker arm, making tappet adjustment difficult.

7   Check the free length of each valve spring and replace the whole set if any has taken a permanent set. Worn or 'tired' valve springs have a marked effect on engine performance and should preferably be renewed during each decoke as a minimum, especially in view of their low overall cost.

8   The valve guides are an interference fit in the cylinder head, and should be drifted out from the combustion chamber side, using a double diameter drift. Before attempting to move the guides, clean off any deposits of carbon from the portion of guide which projects into the valve port. This will ease removal and prevent damage to the cylinder head. New valve guides should be fitted in a similar manner, from the opposite side of the cylinder head, and then reamed out to give the specified valve guide (stem clearance).

**Note:** Before removal or installation of valve guides is undertaken, the cylinder head must be heated to 150°C (300°F) to expand the head and so aid fitting. Unless the cylinder head is heated evenly, distortion may occur and for this reason heating should take place in an oven. Generally, it is best to seek specialist advice when renewing valve guides, as there is always risk of distorting the cylinder head casting.

9   After fitting new valve guides the valve seats must be recut to restore concentricity of the seat with the guide bore. Grinding should then be carried out, as described in paragraph 3.

10  Check the sealing of valve by inverting the cylinder head and pouring paraffin or petrol into the combustion chamber. Allow the head to stand for some time and check that no liquid has seeped into either port.

19.2 Mark 'C' on piston indicates cylinder/piston classification

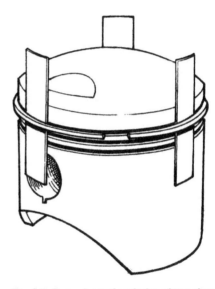

Fig. 1.4. Removing and replacing piston rings

21.1a Use compressor to displace valve collets

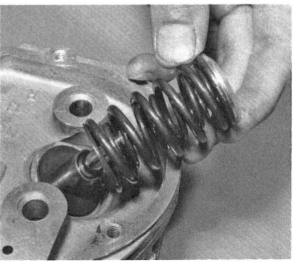

21.1b Remove the springs and ...

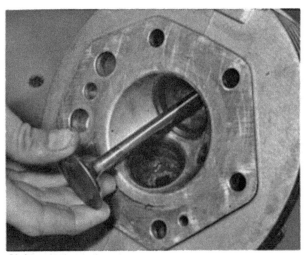

21.1c ... push out the valve

## 22 Cylinder heads: examination and renovation

1   Remove all traces of carbon from the combustion chambers and the inlet and exhaust ports, using a soft scraper which will not damage the surface of the valve seats. Finish by polishing the combustion chamber and ports with metal polish so that carbon does not adhere so readily. Never use emery cloth since the particles of abrasive will become embedded in the soft metal.
2   Check to make sure the valve guides are free from oil or other foreign matter that may cause the valves to stick.
3   If the valve seats are pocketed, as the result of excessive valve grinding in the past, the valve seats should be re-inserted. This is a specialist task which requires expert attention and is quite beyond the means of the average owner. Pocketed valves cause a marked fall-off in performance and reduced engine efficiency as a direct result of the disturbed gas flow.
4   Make sure the cylinder head fins are not clogged with oil or road dirt, otherwise the engine may overheat. If necessary, use a wire brush but take care not to damage the light alloy fins.
5   Check that there are no cracks, and that the valve guides are secure.

## 23 Rocker assemblies: examination and renovation

1   Examine carefully the outer surfaces of each rocker arm, to ensure there are no surface cracks or other signs of premature failure. The rocker arms should have a smooth surface to resist any tendency towards fatigue failure.
2   The rocker arms should be a good sliding fit on the rocker spindles without excessive play. Noisy valve gear will result from worn rocker arms and spindles and performance may drop off as a result of reduced valve lift. If play is evident, the rocker arms should be renewed and new spindles fitted.
3   Check the rocker arm adjuster and the end of the rocker which engages with the pushrod. Both these points of contact have hardened ends and it is important that the surface is not scuffed, chipped or broken, otherwise rapid wear will occur.
4   The spacer on the rocker spindle should revolve freely without endfloat.

## 24 Engine reassembly: general

1   Before reassembly, the various engine components should be thoroughly cleaned and laid out close to the working area.
2   Gather together all the necessary tools and have available an oil can filled with clean engine oil. Make sure all the new gaskets and oil seals are to hand, also any replacement parts required. There is nothing more infuriating than having to stop in the

middle of a reassembly sequence because a vital gasket or replacement part has been overlooked.
3   Make sure the reassembly area is clean and well lit and that there is adequate working space. Refer to the torque and clearance settings, wherever they are given. Many of the smaller bolts are easily sheared if they are overtightened. Always use the correct size spanner and screwdriver, never an adjustable or grips as a substitute. If some of the nuts and bolts that have to be replaced were damaged during the dismantling operation, renew them. This will make any subsequent reassembly and dismantling much easier.
4   Above all else, use good quality tools and work at a steady pace, taking care that no part of a reassembly sequence is omitted. Short cuts invariably give rise to problems, some of which may not be apparent until a much later stage.

## 25 Engine reassembly: replacing the crankshaft and connecting rods

1   If the front main bearing housing was removed, it should be refitted first. **Ensure that the housing is replaced so that the oil ways align.**
2   Place the engine, front downwards, with the casing well clear of the workbench. Lubricate the front main bearing thoroughly and insert the crankshaft. Fit the oil seal to the rear main bearing housing with its spring side facing inwards. Note that the seal should only be inserted far enough for its outer face to lie flush with the seal housing. Lubricate the bearing and the seal lips, and refit the main bearing housing. The main bearing oil return pipe projecting from the housing must pass through the hole in the crankcase wall, and the oil ways between the housing and crankcase must line up. After tightening the housing bolts do not omit to bend up the ears of the locking plates. Check that the crankshaft rotates easily.
3   Reposition the engine so that it is supported on the cylinder head studs. Fit the big-end shells to the connecting rods and end caps, ensuring that the locating tongues on the shells locate correctly with the recesses in the holders. Lubricate the big-end journal and refit the connecting rods and caps. **The connecting rods must be fitted in their original positions on the big-end journals.** The oil hole in the left-hand connecting rod must face upwards (with the engine in the normal position) and that in the right-hand connecting rod downwards. In addition, the milled edges of the connecting rods and caps must all be on the same side. Tighten the big-end nuts to a torque setting of 4.6 - 4.8 kg m (33 - 35 ft lb).
4   Check again for easy rotation of the crankshaft. Binding of the big-end bearings can be caused by incorrect selection or displaced shells.
5   Whilst the engine is in this position the sump may be refitted. Before replacing the sump the oil strainer should be cleaned and

25.1 Crankshaft plug must be absolutely tight

25.2a Insert the crankshaft and ...

25.2b ... refit the rear main bearing housing and gasket

25.2c Do not omit the push fit pipe

25.3a Fit the big end bearing shells and ...

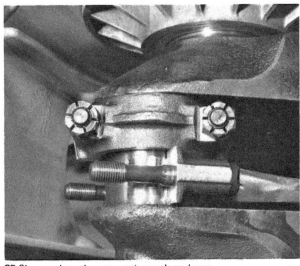

25.3b ... replace the connecting rods and caps

25.5 Refit the sump onto a new gasket

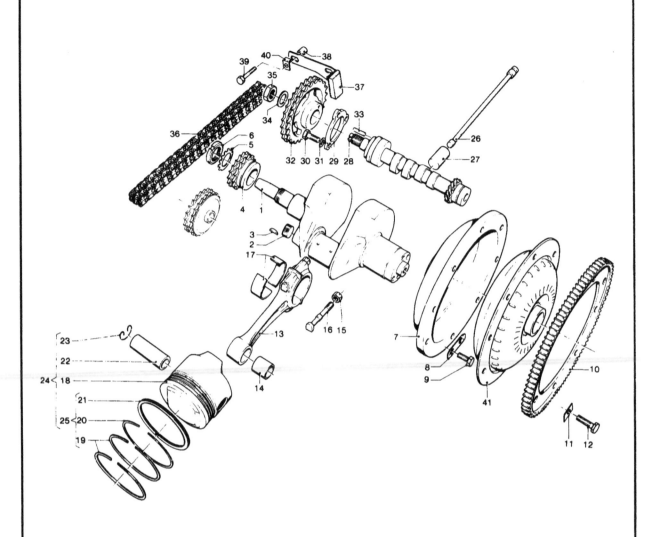

**Fig. 1.5. Crankshaft and timing chest components**

| | | | | | | |
|---|---|---|---|---|---|
| 1 | Crankshaft | 15 | Self locking nut - 4 off | 28 | Camshaft |
| 2 | Plug | 16 | Big-end bolt - 4 off | 29 | Flange |
| 3 | Woodruff key | 17 | Big-end bearing shell set - 2 off | 30 | Bolt - 3 off |
| 4 | Crankshaft sprocket | 18 | Piston - 2 off | 31 | Spring washer - 3 off |
| 5 | Tab washer | 19 | 1st and 2nd compression ring - 4 off | 32 | Camshaft sprocket |
| 6 | Peg nut | 20 | 3rd compression ring - 2 off | 33 | Drive pin |
| 7 | Flywheel - V-1000 type | 21 | Oil control ring - 2 off | 34 | Washer |
| 8 | Lock plate - 3 off | 22 | Gudgeon pin - 2 off | 35 | Nut |
| 9 | Bolt - 6 off | 23 | Circlip - 4 off | 36 | Cam chain |
| 10 | Starter ring gear - V-1000 type | 24 | Piston complete - 2 off | 37 | Chain tensioner |
| 11 | Tab washer - 4 off | 25 | Piston ring set - 25 off | 38 | Spacer - 2 off |
| 12 | Bolt - 4 off | 26 | Pushrod - 4 off | 39 | Bolt - 2 off |
| 13 | Connecting rod - 2 off | 27 | Cam follower - 4 off | 40 | Tab washer - 2 off |
| 14 | Smal-end bush - 2 off | | | 41 | Torque converter - V-1000 only |

the oil filter (if utilised) renewed. In addition, the opportunity should be taken to dismantle and clean the oil pressure control valve. See Chapter 3.11 for details of the procedure.

### 26 Engine reassembly: replacing the camshaft, oil pump and timing the valves

1   Place the timing chain tensioner arm in position in the casing with the distance piece on each bolt between the arm and the casing. Tighten the bolts lightly.

2   Lubricate the camshaft journal and insert the camshaft into the crankcase. Fit the endplate and tighten the three screws.

3   Position the oil pump in the casing and fit the four retaining bolts. The oil pump must be fitted so that the drive spindle is offset from the line of the crankshaft and camshaft. Tighten the four bolts evenly to 3 kg m (22 ft lbs), checking by rotation of the driveshaft that the pump gears do not bind.

4   Insert the drive pin into the eccentrically drilled hole in the camshaft and replace the Woodruff keys into the keyways of the other two shafts. Arrange the camshaft, crankshaft and oil pump sprockets on the workbench and fit the cam chain as it would be fitted on the engine. The timing mark on the camshaft sprocket should be aligned roughly with that of the crankshaft sprocket. Lift the complete assembly up and engage each sprocket with its respective shaft, holding the two sprockets in alignment.

5   Rotate the camshaft and crankshaft until the keys align with the keyways in the sprockets. Push the sprockets fully home on the shafts. The index marks on the two upper sprockets **must be in alignment.** One or two attempts may have to be made to ensure the timing is accurate.

6   Replace the sprocket retaining nuts and washers. The tab washer on the special crankshaft nut is inordinately difficult to bend up after the nut has been tightened, due to lack of space. It is worth bending up the extreme ends of the ears before refitting, to aid final securing of the nut.

7   As when loosening the nuts, rotation of the shafts may be prevented by passing a close fitting bar through the small end eye, resting on wooden blocks across the crankcase mouth.

8   Push the chain tensioner across towards the middle of the casing until all slack has been taken up. Hold the tensioner arm in the chosen position and tighten the two bolts. Rotate the crankshaft a number of times and check that the chain does not become overtight at any point. If this occurs, slacken the tensioner bolts and readjust at the tightest point. Secure the two bolts by means of their locking plates.

9   Place a new gasket on the timing chest mating surface and refit the cover. On V-1000 models the hexagonal drive piece fitted into the end of the camshaft must engage correctly with the converter oil pump drive boss. Fit the cover bolts and tighten them evenly, in a diagonal sequence, to avoid distortion.

10  When refitting the timing cover, lubricate the lip of the crankshaft oil seal.

26.1 Position the chain tensioner and tighten the screw tightly

26.2a Insert the camshaft and ...

26.2b ... replace the end plate and screws

26.3 Place the oil pump in place, located by dowels

26.4a Replace the oil pump shaft and crankshaft Woodruff keys and ...

26.4b ... insert the camshaft drive pin

26.5 Timing marks must line up

26.6 Do not omit crankshaft nut tab washer

26.9 Lubricate oil seal lip ...

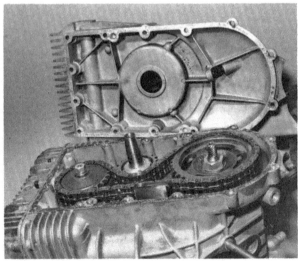

26.10 ... before fitting timing cover onto crankcase

## 27 Engine reassembly: replacing the pistons, cylinder barrels and cylinder heads

1   As with dismantling, the cylinder assemblies should be refitted individually, taking care that components from one assembly do not become interchanged with those of another. Care should be taken also that matched components are fitted in their original positions.

2   Lubricate and refit all four cam followers in their original positions.

3   Fit the piston rings to the piston of the cylinder being attended to. Fit the oil control ring from the skirt side and each of the compression rings from the crown side, commencing with the third ring. Thin strips of metal may be used to aid ring refitting and reduce the risk of breakage.

4   Lubricate the small end bush with clean engine oil. Refit one wire circlip to the piston. Warm the piston in boiling water and press the gudgeon pin fully home through the piston bosses and small end eye. The arrow cast on the piston crown must face forwards. Refit the second circlip and check that both are correctly seated in their locating grooves.

5   Position a new cylinder base gasket over the holding down studs, **ensuring that the oil hole in the gasket aligns with the oil hole in the cylinder seat mating surface.** Place a new 'O' ring on each of the two short holding down studs.

6   Refitting of the cylinder barrel is facilitated if the piston is supported squarely at TDC, on two wooden blocks. Lubricate the cylinder bore and position the ring gaps at 120° relative to one another. Place the cylinder barrel over the piston crown and compress each ring individually so that the barrel may be pushed home. A piston ring clamp will aid refitting. Remove the two support blocks and seat the cylinder barrel against the gasket.

7   Before refitting the cylinder head, the valves and springs must be installed by reversing the dismantling procedure. Do not omit the shims and washers which lie below the lower spring seat. Ensure that the two split collets engage correctly with the groove in the valve stem. After releasing the spring compressor, seat the valve collars by striking squarely on the end of the valve stem.

8   Place a new cylinder head gasket over the holding down studs. As with the cylinder base gasket, **it is important that the oil hole lines up with that of the cylinder barrel.** Place the cylinder head in position and fit an 'O' ring to each of the four long studs. Replace the rocker arm support bracket so that the spindle screw holes are at the top.

9   Refit the six cylinder head nuts and washers. The smaller wave washer fits the upper short stud. Tighten the cylinder head nuts down evenly, in a diagonal sequence, to a torque setting of 4 - 4.5 kg m (29 - 32 ft lb).

10 Insert both pushrods into the tunnel in the cylinder head so that the ball ends of each pushrod engage with the cam follower cups. Rotate the engine until the piston is at TDC, with both valves closed. Replace both rocker arms, bronze washers and double coil washers. The bronze washer should lie between the spring washer and support bracket. Insert the rocker spindles and, using a screwdriver in the slotted lower ends, rotate each spindle until the locating screws can be inserted and tightened.

11 Refit the rocker oil feed pipe to the cylinder head. If necessary, use two new sealing washers at the banjo union.

12 Before replacing the rocker cover, the valve clearances should be adjusted, using a feeler gauge. Refer to the Specifications for the correct clearance. Adjust the gap by means of the adjuster screw after slackening the locknut, with the piston at TDC on the compression stroke.

13 Replace the rocker cover together with a new gasket. Note that on some models the rearmost inner cover screw on the left-hand cylinder also retains the choke lever assembly.

14 Repeat the assembly procedure for the second cylinder.

## 28 Engine reassembly: replacing the flywheel and clutch (except V-1000 Convert model)

1   Place the flywheel on the crankshaft boss, rotating it until the TDC mark on the flywheel is aligned with the white index mark on the boss. This will ensure that the timing marks on the flywheel periphery are in correct relationship with the crankshaft. Insert the flywheel retaining bolts together with the shared lock plates. Tighten the bolts evenly and then secure them by bending up the ears of the plates.

2   Install each of the eight clutch springs in the recesses in the flywheel face. Apply a small amount of heavy graphite grease to the internal splines in the flywheel. Apply the grease sparingly to prevent any finding its way onto the clutch plates. Insert the clutch spring plate into position so that the index mark on the plate aligns with the TDC mark on the flywheel. This will aid alignment of the springs with the recesses in the rear of the spring plate.

3   Replace the two friction plates and the intermediate plate, followed by the starter ring gear. Insert and start the eight bolts, but do not tighten them at this stage. Because the two friction plates are fully floating when not under spring pressure they must be centralised before tightening down the starter ring gear, to enable easy entry of the splined boss of the gearbox input shaft. A special tool is recommended with which to accomplish this task. If this tool is not available the splined boss may be removed from the gearbox input shaft and used as the centraliser. When using the latter method, first insert the clutch thrust piece to aid centralisation of the boss.

4   With the tool in place tighten down the eight bolts evenly, about one turn at a time against the spring pressure. The centralising tool should be an easy sliding fit if the plates are positioned correctly. Remove the tool and insert the clutch thrust piece, if this has not been done already.

27.2 Lubricate and fit cam followers in original positions

27.4a Arrow on piston must face forwards

27.4b Insert gudgeon pin fully and ...

27.4c ... refit the circlip ensuring it seats correctly.

27.5 Fit new cylinder base gasket and 'O' ring

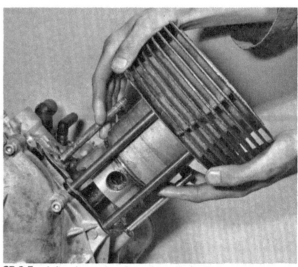
27.6 Feed the piston rings into the cylinder bore

27.7 DO NOT omit correct number of shims

27.8a Fit cylinder head onto new gasket

27.8b Place new 'O' rings on long stud

27.8c Replace the rocker support bracket

27.8d Small washer fits on short upper stud below ...

27.9 ... the special sleeve nut

27.10a Insert both pushrods , cupped ends uppermost

27.10b Replace rocker arm, shim and spring washer and ...

27.10c ... insert the spindle, so that hole aligns to take screw

27.10d Do not omit internal cap

27.11a Check valve clearances before replacing cover

27.11b Use new sealing washers at upper and ...

27.11c ... lower unions on rocker feed pipes

28.1a Refit flywheel so that white mark aligns with TDC mark on flywheel ...

28.1b Note the locking device, used when tightening the bolts

### 29 Engine reassembly: replacing the flywheel and torque converter (V-1000 Convert model only)

1   Place the engine in the same position as that used during removal of these components. Refit the flywheel in the manner described for other models in the previous Section.
2   Position the starter ring gear and the torque converter on the flywheel and insert and tighten the bolts lightly. It is essential that the boss of the torque converter into which the gearbox input shaft fits, runs absolutely concentrically. If this precaution is not taken, difficulty may be encountered when trying to fit the gearbox to the engine. Additionally, oil leakage will occur at the oil seal. To check for correct location, a dial gauge or index gauge should be mounted on a bracket retained on one of the gearbox mounting studs. If eccentricity is apparent, remove the four retaining bolts and move the torque converter one quarter of a turn until the bolt holes again line up. Refit the bolts and check again. The maximum permissible run out measured on the outer surface of the boss is 0.05 - 0.06 mm (0.0019 - 0.0023 in).
3   When the torque converter is running true, tighten the bolts and secure them by means of the locking tabs.

### 30 Engine reassembly: joining the engine to the gearbox

1   Fitting of the gearbox onto the engine is straightforward on both manual and automatic models. If difficulty is encountered when inserting the splined boss into the centre of the clutch on manual gearbox models, a long-handled screwdriver may be inserted between the two mating surfaces, with which to rotate the flywheel.
2   The splined boss should be lubricated with a small quantity of graphite grease before assembly.
3   On V-1000 models, lubricate the torque converter boss with clean hydraulic fluid before fitting. It is not necessary to replenish the torque converter with hydraulic fluid at this juncture. The bush on the end of the input shaft must be fitted with the slotted end outwards, ie, towards the engine.
4   After pushing the gearbox fully home, fit and tighten the retaining nuts and washers.

### 31 Engine reassembly: replacing the distributor and alternator

1   If during dismantling the relative positions of the distributor, contact breaker base plate and the contact breaker cam were

scribe marked or punch marked, replacement of the distributor in the correct position is straightforward. Rotate the engine until the right-hand piston is at TDC on the compression stroke. Turn the contact breaker cam until the chosen mark aligns with that made on the body. Holding the cam in this position insert the distributor into the crankcase, so that the driven gear engages with the gear on the camshaft and the distributor and crankcase marks line up. Fit and temporarily tighten the distributor retaining clamp. The contact breaker gaps can now be set and the static ignition timing carried out as described in Chapter 4, Section 7.
2   Where index marks were not made during dismantling, the distributor may be refitted as an integral part of the static ignition timing procedure. The distributor should be inserted when the right-hand cylinder timing mark is visible in the crankcase aperture, and when the right-hand cylinder contact breaker points are just on the verge of opening. The distributor must be refitted with the direction indicating arrow on the distributor boss facing away from the right-hand cylinder. If this precaution is not taken, the standard ignition timing procedure will have to be rearranged.
3   Lubricate the alternator rotor boss so that on replacement, the boss may enter the timing cover oil seal more easily. Fit the rotor on the shaft and tighten the retaining socket screw. Position the stator over the rotor so that the multi-pin terminal is on the right. Align the bolt holes in the stator with those in the casing and push the stator fully home into the casing recess. Fit and tighten evenly the three socket screws. Free the brushes so that they contact the slip rings, under pressure from their springs.

### 32 Engine reassembly: refitting the engine in the frame

1   Before removing the engine/gearbox unit from the workbench, refit the two sub-frame members either side of the engine, together with the centrestand. At this stage the sub-frame should be secured by the gearbox mounting bolt.
2   Position the engine/gearbox unit on the floor, supported on the same blocks used when dismantling was being carried out. Ensure that the complete assembly is secure. Where utilised, the air filter box should be placed between the cylinders in approximately the correct position before the frame is lowered into place.
3   Wheel the frame assembly into position and lower it into place so that the sub-frame to main frame joining lugs are in approximately the correct relationship. It may be necessary to spring the two forward frame tubes outwards to clear the sub-frame lugs. Insert the two short socket bolts and fit the nuts loosely. On touring models the front crash bars must be fitted with these bolts. Insert the front engine mounting bolt and fit the nut. On non-touring models the bolt must be fitted from the left-hand side, together with the propstand. Ensure that the projection on the propstand bracket engages with the recess in the frame lug. **Do not** tighten any frame bolts at this stage.
4   Replace the rear swinging arm unit and the rear wheel by following the procedure described in Chapter 5, Section 8.
5   Position the battery carrier plate on the gearbox and fit and tighten the retaining bolts. **Do not** omit the battery earth strap which is secured by the left-hand rear bolt.
6   Fit and tighten the four sub-frame rear bolts. On touring machines the rear bolts retain the silencer support brackets. On other models the bolts secure the footrest bars. Now that all the sub-frame to main frame bolts have been refitted, they may all be tightened fully.

### 33 Engine reassembly: completion

1   Refit the starter motor and connect the main leads to the solenoid. Reconnect also the neutral indicator wire at the switch (where utilised) and the wires at the remaining indicator switches. The wires running to the coils from the contact breaker set may also be replaced. Refer to the wiring diagrams if any doubt exists as to the correct wire positions.

32.3a Refit the front engine mounting bolt and ...

32.3b ... replace the front clamping socket screws

32.3c Note the position of the spacers on gearbox bolt

2   Insert a new air filter element (where utilised) into the air box and fit the breather box so that the central screw projects through the box end. Note that the flange plate on the breather box is located in one position by a projection and recess. Fit and tighten the end nut. If difficulty is encountered in fitting the nut and washer, detach the support bracket from the frame and then fit the nut and washer. The bracket can then be replaced on the frame tube with ease. Reconnect the four breather pipes which locate with individual unions on the breather box. On models not utilising an air filter, refit and reconnect the breather box.

3   On models fitted with an air filter box, replace the rubber duct and secure it in position with the strap and spring arrangement. Replace both carburettors, using new flange gaskets if the inlet stubs were separated from the cylinder heads. Note that the lower bolt securing each inlet stub also retains the HT lead guide clip. Reconnect the throttle cables to the throttle slides and carburettor tops. Fit the throttle return springs and replace the respective assemblies in the carburettors. Where cable controlled chokes are utilised, replace the control lever on the left-hand rocker cover and insert each choke assembly into the carburettors. Pass the petrol feed pipe assembly under the carburettors and reconnect the pipe to the carburettors. Do not omit the filter screens.

4   Replace the exhaust system using a new gasket ring at each port. The slot in each split port collar should face downwards. **Do not** tighten the exhaust flange nuts fully. Tighten the two nuts evenly until the gasket can be felt to be compressed slightly. If leakage occurs when the engine is run, the nuts may be tightened a little further.

5   Reconnect the leads running to the alternator. Refer to the relevant wiring diagram for the correct positioning of the wires. The three wire sockets should be refitted with the small projection on the rubber cover facing inwards. Engage the wiring grommet and wires in the recess in the top of the casing and replace the alternator cover and screws.

6   Refit the gearchange and rear brake link rods and levers. Use new split pins to secure the clevis pins. On touring models, the footboards should be refitted, to aid adjustment of the lever operating angles.

7   On V-1000 models, reconnect all the torque converter fluid system pipes. Use new banjo union sealing washers if there is any doubt as to their condition. Check that all the unions are secure and that the pipes are tracked correctly and not   pinched at any point.

8   Reconnect the speedometer cable and tachometer cable (where utilised). Using a suitable lever operate the clutch release arm at the gearbox and reconnect the cable. **Do not** omit to replace the starter cut-out switch leads.

9   Refit the petrol tank and reconnect the pipes, securing them by means of the clips. Reconnect the electro-valve wire (where fitted) and the petrol level indicator wire on V-1000 models.

10 Replenish the engine with the correct quantity and grade of oil, through the filler orifice in the left-hand side of the crankcase. If the gearbox was overhauled, this too should be replenished with lubricant. On V-1000 models, refill the  torque converter fluid reservoir with the correct specification hydraulic fluid. When the engine is started the fluid level will drop considerably as the torque converter system is progressively filled. The quantity should be augmented by the addition of more fluid as the level falls. When the level has stabilised, refill to the upper level mark.

11 Replace the battery and reconnect the leads. The red lead must be connected to the positive (+) terminal and the black lead to the negative (−) terminal.

## 34 Starting and running the rebuilt engine

1   With the battery negative lead reconnected, check that all electrical accessories work. Do not operate the starter yet.

2   Go through each step of the reassembly procedure and

satisfy yourself that every task has been carried out correctly and all nuts or bolts have been tightened.

3   Turn the engine over by means of the rear wheel, with the gearbox in top gear, to make sure that it rotates freely.

4   Replace the spark plugs, correctly gapped and the spark plug caps.

5   Switch on the ignition and petrol and check that the oil and charge warning lights come on. Operate the starter and run the engine at fast tickover. If the warning lights do not go out, switch off immediately and investigate.

6   The exhaust will smoke considerably at first, due to oil present from reassembly. The smoke should clear gradually.

7   If the engine refuses to start, check that petrol is entering the carburettors, and that there is a spark at the plugs.

8   Check the timing with a stroboscope (see Chapter 4, Section 7).

9   Make sure that there are no oil leaks. Slight seepage can often be cured by further tightening when the engine has bedded down again.

10 Before taking the machine on the road, check that all controls operate freely, and both brakes are in adjustment. Also check that all housings have been refilled with oil.

11 If the engine has been rebored, or if a number of new parts have been fitted, a certain amount of running-in will be required. Particular care should be taken during the first 100 miles or so, when the engine is most likely to tighten up, if it is overstressed. Commence by making maximum use of the gearbox or in the case of the V-1000 model, not allowing the engine to labour so that only a light loading is applied to the engine. Speeds can be worked up gradually until full performance is obtained with increasing mileage. It is particularly important that chromed bores are run in very carefully. If this is not done, flaking of the chrome may result, necessitating renewal once more.

12 After running-in, the engine should be serviced, adjustments checked and bolts and nuts tightened. Drain and refill the sump.

33.4a Insert a new ring gasket into each port

33.4b The gap in the collar must face inwards

33.4c Tighten the flange bolts evenly to avoid distortion

33.4d Ensure that the silencer bolts are scure

**35 Fault diagnosis: engine**

| Symptom | Cause | Remedy |
|---|---|---|
| Engine will not start | Contact breaker points closed or dirty | Check and readjust points. |
| | Flooded carburettor | Check whether float needle is sticking and clean. |
| | Loose or faulty HT cable | Check. |
| | Stuck valve | Clean valve stem. |
| Engine runs unevenly and misfires | Incorrect ignition timing | Check setting and adjust if necessary. |
| | Faulty or incorrect grades of spark plug | Clean or replace plugs. |
| | Fuel starvation | Check fuel lines and carburettor. |
| | Valve clearance too small | Readjust. |
| | Leaky valve | Regrind valve seat. |
| | Low compression | Check and rebore or replace cylinder barrels and rings. |
| Lack of power | Incorrect ignition timing (retarded) | Check and reset timing. Check action of automatic advance unit. |
| Engine pinks | Incorrect ignition timing (over-advanced) | Check and reset timing. |
| Excessive mechanical noise | Worn cylinder barrels (piston slap) | Rebore and fit O/S pistons, or renew cylinder barrels. |
| | Worn small end bearings (rattle) | Replace bearings, gudgeon pins and pistons. |
| | Worn big-end bearings (knock) | Replace shell bearings and regrind crankshaft. |
| | Worn main bearings (rumble) | Fit new bearings. |
| Engine overheats and fades | Lubrication failure | Check oil pump and oil pump drive. |
| Engine overheats | Timing retarded | Readjust. |
| | Incorrect spark plugs | Renew. |

# Chapter 2 Gearbox, clutch and torque converter

## Contents

## Specifications

### V-1000 I-Convert

#### Gearbox

| | |
|---|---|
| Type ... ... ... ... ... ... ... ... ... | Three shaft, constant mesh |
| No. of ratios ... ... ... ... ... ... .. . ... | Two |
| Ratios: | |
| 1st gear ... ... ... ... ... ... ... ... | 1.333 : 1 |
| 2nd gear ... ... ... ... ... ... ... ... | 1.000 : 1 |
| Primary ratio ... ... ... ... ... ... ... | 1.570 : 1 |
| Mainshaft endfloat ... ... ... ... ... ... | 0.15 - 0.20 mm (0.006 - 0.008 in) |
| Layshaft/bush clearance ... ... ... ... ... | 0.040 - 0.106 mm (0.0015 - 0.0041 in) |
| Bush/pinion clearance ... ... ... ... ... ... | 0.000 - 0.390 mm( Zero - 0.0153 in) |

#### Clutch

| | |
|---|---|
| Type ... ... ... ... ... ... ... ... ... | Dry, multi-plate |
| No. of plates | |
| Plain ... ... ... ... ... ... ... ... | 5 |
| Friction ... ... ... ... ... ... ... | 6 |
| Friction plate thickness ... ... ... ... ... | 3.15 - 3.35 mm (0.124 - 0.131 in) |
| Wear limit ... ... ... ... ... ... ... | 2.65 mm (0.104 in) |
| No. of springs ... ... ... ... ... ... ... | 6 |
| Free length ... ... ... ... ... ... ... | 27.970 - 28.000 mm (1.101 - 1.102 in) |

#### Torque converter

| | |
|---|---|
| Type ... ... ... ... ... ... ... ... | Hydraulic turbine |
| Maximum converting ratio ... ... ... ... ... | 1.6 : 1 |
| Working clearances: | |
| Converter housing/flanged shaft ... ... ... ... ... | 0.010 - 0.059 mm (0.0004 - 0.0020 in) |
| Flanged shaft/converter spigot ... ... ... ... ... | 0.070 - 0.104 mm (0.0027 - 0.004 in) |
| Flanged shaft/ bush outer diameter ... ... ... ... | 0.058 - 0.149 mm (0.0022 - 0.006 in) |
| Clutch shaft/bush inner diameter ... ... ... ... ... | 0.006 - 0.035 mm (0.0002 - 0.0013 in) |
| Converter boss/crankshaft spigot ... ... ... ... ... | 0.016 - 0.043 mm (0.0006 - 0.0016 in) |

**Specifications**

## Converter oil pump

| | |
|---|---|
| Type ... ... ... ... ... ... ... ... ... | Trochoid |
| Working pressure ... ... ... ... ... ... ... | 25 - 30 psi |
| Gearbox oil capacity ... ... ... ... ... ... ... | 0.6 ltr (16.9 Imp oz) |
| Torque converter system oil capacity ... ... ... ... ... | 1.5 - 1.7 ltr (42.9 - 47.8 Imp oz) |

# 750-S3, 850T-3 and Le Mans models

## Gearbox
Ratios:

| | |
|---|---|
| 1st gear ... ... ... ... ... ... ... ... | 11.643 : 1 |
| 2nd gear ... ... ... ... ... ... ... ... | 8.080 : 1 |
| 3rd gear ... ... ... ... ... ... ... ... | 6.095 : 1 |
| 4th gear ... ... ... ... ... ... ... ... | 5.059 : 1 |
| 5th gear ... ... ... ... ... ... ... ... | 4.366 : 1 |
| Primary ratio ... ... ... ... ... ... ... | 1.235 : 1 |

# 750S and 850T models

| Ratios | 750S | 850T |
|---|---|---|
| 1st gear ... ... .. ... ... ... ... ... | 10.806 : 1 | 11.424 : 1 |
| 2nd gear ... ... ... ... ... ... ... ... | 7.499 : 1 | 7.928 : 1 |
| 3rd gear ... ... ... ... ... ... ... ... | 5.657 : 1 | 5.980 : 1 |
| 4th gear ... ... ... ... ... ... ... ... | 4.695 : 1 | 4.963 : 1 |
| 5th gear ... ... ... ... ... ... ... ... | 4.052 : 1 | 4.284 : 1 |
| Primary ratio ... ... ... ... ... ... ... | 1.235 : 1 | 1.235 : 1 |
| Working clearances: | | |
| Mainshaft bearing/cover bearing ... ... ... ... ... | 167.1 - 167.2 mm (6.578 - 6.582 in) | |
| Layshaft adjusted length ... ... ... ... ... ... | 144.7 - 145.2 mm (5.692 - 5.715 in) | |
| Gearbox oil capacity ... ... ... ... ... .. ... | 0.75 ltr (1.75/1.33 US/Imp pints) | |

## Clutch

| | |
|---|---|
| Type ... ... ... ... ... ... ... ... ... | Two plate, diaphragm |
| Friction plate: | |
| standard thickness ... ... ... ... ... ... ... | 8 mm (0.3149 in) |
| wear limit ... ... ... ... ... ... ... ... | 7.5 mm (0.2953 in) |

## Torque wrench settings

| | |
|---|---|
| Oil level plug ... ... ... ... ... ... .. ... | 2 kg m (14 lb ft) |
| Cover screw ... ... ... ... ... ... ... ... | 1 kg m (7 lb ft) |
| Mainshaft (output shaft) nut at gearbox end cover ... ... ... | 16 - 18 kg m (115 - 129 lb ft) |

## 1  General description

All the models covered in this manual, with the exception of the V-1000 automatic, utilise a 5-speed constant mesh gearbox driven from the engine via a two-plate flywheel mounted clutch. The gearbox incorporates three shafts, of which the input shaft combines a single reduction gear pinion and a spring-loaded cam-type shock absorber. The intermediate or layshaft has one free running gear pinion only, the remaining four being integral with the shaft. The mainshaft or output comprises five gear pinions and two sliding dog clutches by which means the constantly meshed pinions are engaged. With the exception of the mainshaft 5th gear pinion, which runs directly on the shaft, all pinions are supported on caged needle roller bearings, with independent inner races. The gear shafts themselves run in journal ball or caged needle roller bearings. The clutch consists of two friction plates, one placed either side of an intermediate plate, the complete assembly being secured under pressure from eight coil springs by the starter ring gear bolted to the flywheel.

The V-1000 model is fitted with an unusual transmission system which virtually eliminates the need for gear selection whilst on the move. An hydraulic torque converter transmits the engine power through a traditional multi-plate clutch to a two-speed gearbox. The torque converter, which is oil filled to provide the drive medium, is fitted with a circulatory cooling system comprising an engine driven oil pump and an oil cooler, the latter of which prevents heat build-up in the torque converter from becoming excessive. The gearbox, in common with the 5-speed unit, incorporates three shafts, of which the output or mainshaft carries the selector, dog and free running gears. The gear pinions run on plain bronze bushes.

On all models, helically cut gear pinions are utilised to improve wear characteristics and to reduce gear noise. A selection of straight-cut gears, however, is available for fitting to the Le Mans model, for production racing purposes.

## 2  Gearbox: removal from the frame

1   The gearbox can be removed from the frame only as an integral part of the engine. Separation of the engine from the gearbox is possible only after the frame has been lifted clear. Gearbox removal therefore follows substantially the same procedure as that described for engine/gearbox removal in Chapter 1, Section 4.

2   Separation of the gearbox from the engine should be carried out as described in Chapter 1, Section 6. Drain the oil from the gearbox before commencing the dismantling operations.

2.1 Remove filler plug and drain plug from gearbox

2.2 Separate gearbox after removing bellhousing nuts

## 3  Gearbox: dismantling, 5-speed gearbox only

1  Place the gearbox on the workbench so that the rear end is facing upwards. Remove the clutch operating arm after detaching the clevis pin, which is secured by a split pin. Lift the return spring from position. Withdraw the clutch pushrod thrust components from the gearbox end casing, noting carefully the sequence of the five individual parts. Push the clutch pushrod out and note and displace the small rubber bush from the centre of the input shaft. Remove the gear selector arm, which is retained by a pinch bolt.

2  Invert the gearbox and place it on the workbench, supported on a number of wooden blocks. The splined boss on the end of the input shaft is retained by a special ring nut, secured by a tab washer. The nut must be loosened by means of a peg spanner. This may be fabricated from a length of thick walled tube of suitable dimensions, one end of which has been relieved to form four short pegs. To prevent the shaft rotating when loosening the nut, a small metal sprag must be made which will engage with the splined boss and bear against a block of wood in the casing. A small amount of experimentation will elicit the ideal shape for the sprag. The splined boss is a tight fit on the splines of the input shaft and may require drawing from position. A two or three legged sprocket puller may be used to extract the boss. Alternatively, levers may be used, if care is taken.

3  Invert the gearbox again so that the end cover is once more uppermost. Unscrew the speedometer driveshaft housing and withdraw it from the casing, together with the driveshaft. A small endfloat shim is fitted to the lower end of the driveshaft; this is easily overlooked.

4  Using a box spanner, loosen and remove the shouldered nut from the output shaft. The shaft may be prevented from rotating by fitting temporarily the internally splined sleeve which interconnects the final driveshaft with the rear wheel bevel box input shaft. A strap or chain wrench should be employed with which to hold the sleeve. After removal of the nut, access is made to the speedometer drive worm gear. The gear is driven by the output shaft, drive being given through a single steel ball which engages on the shaft splines and the gear. Remove the steel ball and store it in a safe place.

5  Remove the gearbox end cover retaining socket screws. Separate the cover from the gasket, using a rawhide mallet, and lift it clear of the shafts.

6  Lift the shims and washers off the end of the gear selector drum, noting their number and sequence. Unscrew the gearbox breather union from the left-hand wall and withdraw the spring

and detent plunger from the tunnel. The breather union serves also as the plunger housing bolt. Remove the neutral indicator switch as a complete unit by unscrewing the two outer screws which pass through the switch flange.

7  Push the top of the selector drum across slightly towards the casing wall, so that the uppermost selector fork and dog clutch may be lifted off the layshaft end. Pull out the rod upon which the selector drum pivots and move the drum to one side to clear the remaining two selector forks. Lift the drum out of the casing, followed by the endfloat shim(s). Note the number of shims. Lift the selector fork rod up so that the rod lower end leaves the casing and lift the rod, complete with the selector forks, out of the gearbox.

8  Grasp the end of the layshaft and also that of the output shaft. The two shafts may be withdrawn from the gearbox simultaneously, complete with the gear pinions. Note the needle thrust bearing, washers and shim which lie between the layshaft lower end and the gearbox wall. Lift the thrust assembly from position.

9  The remaining components within the gearbox, which form the input or reduction shaft, can be driven from position in the casing using a rawhide mallet.

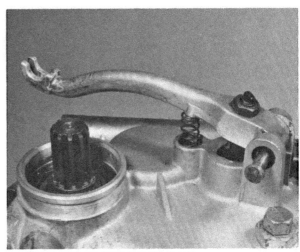

3.1a Displace clevis pin to free clutch arm

3.1b Lift return spring from position

3.1c Remove the pushrod thrust components and ...

3.1d ... withdraw the pushrod and plastic bush

3.2a Unscrew boss peg nut with special spanner after ...

3.2b ... knocking down the ears of the tab washer

3.2c Use a fabricated sprag when loosening the nut

3.3 Unscrew the speedometer drive housing

3.4a Use splined sleeve to prevent shaft rotation

3.4b Unscrew the shouldered nut and ...

3.4c ... displace the speedometer drive ball

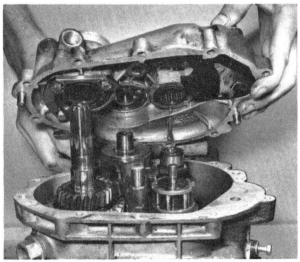
3.5 Lift the end cover off the shafts

3.6a Note and remove change drum shims

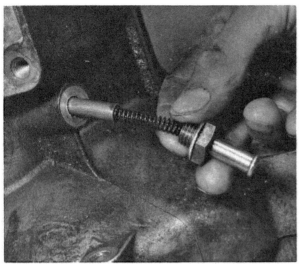
3.6b Unscrew breather and withdraw spring and plunger

3.6c Neutral indicator switch retained by two bolts

3.7a Push drum across to free selector fork

3.7b Pull out shaft to free selector drum

3.7c Mainshaft forks are on one rod

3.8 Lift mainshaft and layshaft out simultaneously

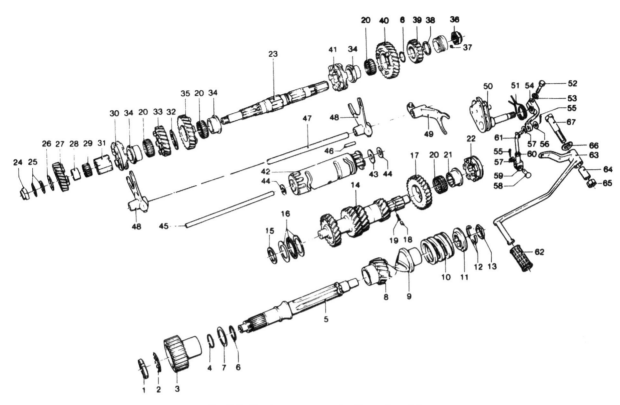

**Fig. 2.1. Gearbox components - 5-speed models**

| | | | | |
|---|---|---|---|---|
| 1 | Peg nut | 18 | Bush locating pin | 35 | Mainshaft 2nd gear pinion | 52 | Pinch bolt |

| | | | |
|---|---|---|---|
| 1 Peg nut | 18 Bush locating pin | 35 Mainshaft 2nd gear pinion | 52 Pinch bolt |
| 2 Tab washer | 19 Spring | 36 Shouldered nut | 53 Star washer |
| 3 Splined input boss | 20 Needle roller bearing | 37 Speedometer gear drive ball | 54 Gearchange arm |
| 4 'O' ring | 21 Inner race | 38 Washer | 55 Split pin - 2 off |
| 5 Input shaft | 22 Sliding dog clutch | 39 Mainshaft 5th gear pinion | 56 Washer |
| 6 'O' ring | 23 Mainshaft (output shaft) | 40 Mainshaft 1st gear pinion | 57 Washer - 2 off |
| 7 Washer | 24 Shouldered nut - LH thread | 41 Sliding dog clutch | 58 Clevis pin |
| 8 Input reduction gear | 25 Shim - as required | 42 Gearchange drum | 59 Clevis fork |
| 9 Shock absorber cam | 26 Washer | 43 Washer | 60 Nut |
| 10 Shock absorber spring | 27 Mainshaft 4th gear pinnion | 44 Shim - as required | 61 Link rod |
| 11 Spring seat | 28 Inner race | 45 Change drum spindle | 62 Rubber |
| 12 Collet - 2 off | 29 Needle roller bearing | 46 Change drum pin - 4 off | 63 Gear lever |
| 13 Spacer | 30 Sliding dog clutch | 47 Selector fork rod | 64 Nylon bush |
| 14 Layshaft | 31 Splined sleeve | 48 Selector fork | 65 Nut |
| 15 Shim | 32 Washer | 49 Selector fork | 66 Wave washer |
| 16 Thrust bearing | 33 Mainshaft 3rd gear pinion | 50 Gearchange selector | 67 Bolt |
| 17 Layshaft 5th gear pinion | 34 Inner race - 3 off | 51 Centraliser spring | |

## 4 Gearshafts and pinions: examination and renovation, 5-speed gearbox only

1 Examine the gear pinions to ensure that there are no chipped or broken teeth and that the dogs on the pinion faces are not rounded. Inspect also the condition of the sliding dogs which are operated by the selector forks. If damage to any of these components is evident, the faulty part must be renewed. Renewal of pinions and further inspection of the shafts, splines and bearings require that the pinions be removed from the shafts. Each shaft should be dismantled separately and the relative positions of the components noted very carefully, to aid refitting.

2 Of the five layshaft gears only the 5th gear pinion is floating; the remainder are an integral part of the shaft. The 5th gear pinion is secured on the shaft by a sping-loaded pin resting in a drilling in one of the shaft splineways. Using a pointed instrument, depress the pin against the spring pressure. Rotate the bush either to the right or left and draw the pinion, caged needle roller bearing and inner bush from position. Care should be taken not to allow the pin and spring to fly from place as the inner bush is drawn off.

3 Unscrew the shouldered nut from the end of the mainshaft. Note that this nut has a left-hand thread and must therefore be undone in a **clockwise** direction. The roller bearing cage and inner race is a tight fit on the shaft and will require pulling from position, using a standard two or three-legged sprocket puller. The puller feet may be located behind the 4th gear pinion, which will be drawn off with the bearing. After pulling the bearing from position, note the adjusting shims which lie between the bearing and 4th gear pinion face. Remove the needle roller bearing and inner bush. Remove the selector dog, noting that it can be fitted only one way round, with the relieved dogs facing inwards. The 3rd gear pinion and bearing can now be removed, followed by the thick washer and the 2nd gear pinion and bearing.

Working from the other end of the shaft, remove the 5th gear pinion, the 'O' ring and the 1st gear pinion and bearing. Finally slide off the dog clutch.

4   With the exception of the mainshaft 5th gear pinion, all pinions run on caged needle roller bearings supported on separate inner bushes or races. If up and down play can be detected on any bearing, or the rollers or tracks are pitted, the matched components should be renewed.

5   The gearbox input shaft, on which the primary reduction gear is mounted, incorporates a cam type shock absorber unit. The shock absorber unit is unlikely to give trouble until an extended mileage has been covered, when wear on the cam faces or weakening of the coil spring will necessitate renewal. In order to dismantle the shaft assembly, the heavy square section coil spring must be compressed to allow removal of the spring seat and the split retaining collets. A special compressor must be used to carry out this operation safely. Most motorcycle repair agents have a tool of the correct type, which is used also for compressing rear suspension unit springs. If the cam faces have worn through the hardened layer, or if heavy chattering marks are apparent, both cam pieces should be renewed. Check the spring length against that of a new component. A marked shortening in the free state will indicate the need for renewal.

4.2a Depress pin and turn pinion to right or left

4.2b Do not allow pin and spring to fly out

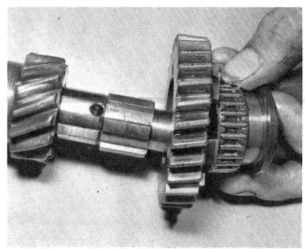

4.2c Pull off the layshaft pinion, caged bearing and race

4.3a Shouldered end nut has LEFT-HAND thread

4.3b Use puller to draw off bearing together with ...

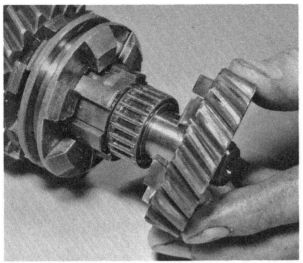
4.3c ... 4th gear pinion and ...

4.3d ... 4th gear pinion bearing

4.3e Remove the sliding dog and ...

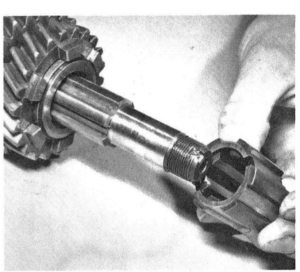
4.3f ... internally and externally splined sleeve

4.3g Displace 3rd gear pinion and bearing and ...

4.3f ... intermediate thick washer

4.3i The 2nd gear pinion and bearing is now free

4.4a Pull the 5th gear pinion off the shaft

4.4b Displace the 'O' ring followed by the ...

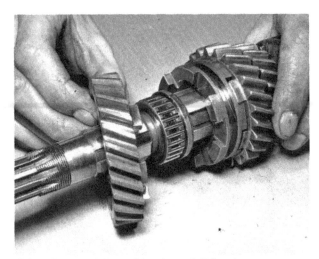

4.4c ... 1st gear pinion and bearing and sliding dog

## 5  Gearselector mechanism: examination and renovation, 5-speed gearbox only

1   The selector forks should be examined closely to ensure that they are not bent or badly worn. The greatest area of wear occurs either side of the fingers which engage with the sliding dogs, and at the pins which locate with the channels in the gear change drum. Check also that the forks are a good sliding fit on the selector rod. Bad gear selection or jumping out of gear can often be traced to the above faults.

2   Check the channels in the change drum for wear. Damage to this component is unlikely unless lubrication failure has occurred. If the change pins in the end of the drum are scored or worn, they may be renewed individually.

3   Ensure that the plunger spring that bears on the change drum has not lost its action and that the pawl end is not excessively worn. Remove the selector pawl assembly from the gearbox end cover and check the condition of the pawl spring and pawl profile. Wear here will be self-evident, as will weakening or fracture of the centraliser spring.

5.1a Push the selector pawl mechanism from end cover

5.1b Check condition of pawls and springs

## 6 Gearbox bearings and oil seals: examination and renewal, 5-speed gearbox only

1   After washing the gearbox bearings thoroughly in petrol or white spirit, check each bearing for roughness when rotated and for up and down play. The bearings should always be checked when still installed in the cases.

2   The bearings may be removed after heating the case to 150 - 160° (300 - 320°F) in an oven. The input shaft ball bearing and the output shaft ball bearing are secured by retainer plates on the inside of the cases. Each plate is retained by three bolts, each of which has a tab washer. After removal of the plates, the bearings may be drifted out. The remaining bearings are fitted into blind holes in the cases and may therefore be difficult to remove. A special expanding puller should be acquired with which to draw them from place. It will be found that a locking fluid was used on many bearings, during original assembly. This will marginally increase the difficulty of removal. When refitting the bearings, the casings should be heated again to the specified temperature. Apply a small quantity of locking fluid to the outer races before inserting them. Attempt to insert each bearing in one operation as the hot cases will dramatically shorten the rate at which the fluid hardens.

3   It is recommended that new oil seals be fitted to the gearbox as a matter of course, during major dismantling. Failure of a re-used oil seal at a later date will require considerable additional dismantling in order to allow renewal. The oil seals may be prised or drifted out of position. Always refit the oil seals with the spring side facing towards the inside of the gearbox. Use a suitable tubular drift to ensure that the seals are not distorted.

## 7 Gearbox reassembly - 5-speed gearbox only

1   If the gear pinion assemblies have been removed from the various shafts, they must be refitted prior to replacing the shafts in the cases. Where new components have been installed which may effect the overall length of the complete gear clusters on each shaft, or if gear selection was difficult and has not been traced to worn components, the overall lengths of the shafts must be adjusted on reassembly by means of shims.

2   Assemble the layshaft components by reversing the dismantling procedure and by referring to the relevant photographs. The total distance should be measured between the points

shown in Fig. 2.3. When taking this measurement a special bronze washer should be fitted in place of the thrust bearing and the two thrust washers which lie either side. This is to give the correct final bearing running clearance. If the distance is incorrect, place shims between the thrust bearing outer washer and the end washer. Both the special bronze washer and shims are available, especially to carry out the operation.

3   A similar procedure should be adopted when refitting the gear pinion assemblies to the mainshaft. The overall length of the completed assembly, measured between the two points shown in Fig. 2.3 should be 144.7 - 145.2 mm (5.6920 - 5.7150 in). Insert additional shims between the mainshaft 4th gear pinion and the inner face of the roller bearing. When replacing the mainshaft end nut, which secures the roller bearing, apply locking fluid to the threads. If locking fluid is not available, bend in a portion of the nut shoulder so that it engages with the short axial channel in the threaded portion.

4   Place the gearbox casing on the workbench supported on blocks. Insert the complete input shaft through the centre bearing and gently drive it home with a rawhide mallet. Do not omit the 'O' ring and the shim from the shaft. Lubricate the bearings in the casing with clean oil. Assemble the two gear shafts together and insert them into the gearbox simultaneously. If difficulty is encountered in holding the layshaft shims and thrust bearing in position, apply a little heavy grease to each shim and the bearing.

5   Slide the two selector forks into position in the two mainshaft sliding dogs. Install the selector drum, together with the shims, into the casing so that the two forks engage with the channels in the drum. Fit the selector drum rod and the selector fork rod. Push the top of the selector drum over towards the gearbox wall, and replace the layshaft selector fork and sliding dog. Rotate the selector drum until the gears are in the neutral position. In this position the two semi-circular cut-outs in the drum pin plate are in the correct position to accept the gearchange pawls when the cover is refitted.

6   The gearbox end cover and gasket should now be replaced temporarily and secured by four diagonally placed socket screws, prior to checking correct gear selection. Insert the speedometer drive gear and the small drive ball, and fit and tighten the shouldered nut. Refit the gear change detent plunger, spring and the breather union cam detent housing.

Adjust the gear selector pawl operation by means of the eccentric bolt located immediately adjacent to the splined gear selector shaft. The eccentric bolt should be placed so that when the shaft is moved slightly to right or left it can be felt that the two pawls are an equal distance away from the pins on the change drum, when the splined shaft is at rest. This procedure should be carried out with the gearbox in the neutral position.

7   Attempt to select each gear in turn a number of times, moving up and down throughout the whole range. Rotation of either the input shaft or the output shaft will aid selection. Assuming that all other areas of gearbox assembly are correct, difficult selection may be due to an incorrectly positioned selector drum. If problems are encountered in selecting 1st gear and 3rd gear, shims should be removed or inserted between the end of the selector drum and the gearbox wall. Difficulty in selecting 2nd gear and 4th gear, may be remedied by placing shims between the selector drum and gearbox end cover. Shims are available in 0.6, 0.8, 1.0 and 1.2 mm (0.023, 0.031, 0.039 and 0.047 in) sizes. It is important that a small amount of selector drum endfloat remains after adjustment.

8   When selection is correct, the gearbox end cover should be secured with the total number of screws. Tighten the output shaft nut to the specified torque setting (see Specifications) and secure it by bending a portion of the nut shoulder into the shaft groove. Refit the neutral indicator switch, with the curved portion of the brass contact strip away from the drum.

9   Lubricate the clutch pushrod and slide the plastic guide over the rod. Insert the rod into the gearbox, ensuring that the guide enters the hollow input shaft. Assemble the clutch thrust

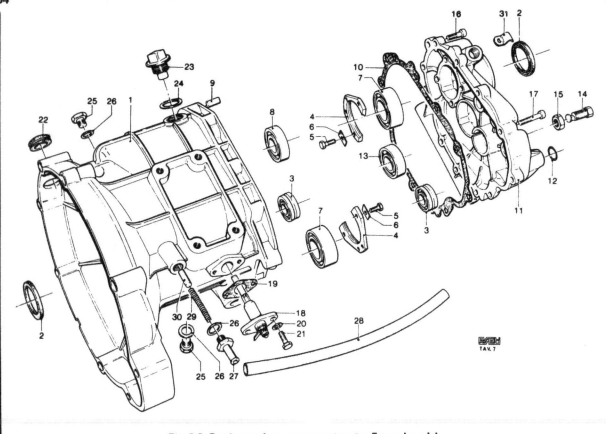

**Fig. 2.2. Gearbox casing - component parts - 5 speed models**

| | | | | | |
|---|---|---|---|---|---|
| 1 | Gearbox housing | 11 | End cover | 21 | Bolt - 2 off |
| 2 | Oil seal - 2 off | 12 | 'O' ring | 22 | Inspection plug |
| 3 | Journal ball bearing - 2 off | 13 | Journal ball bearing | 23 | Oil filler plug |
| 4 | Bearing retainer plate - 2 off | 14 | Eccentric centraliser screw | 24 | Sealing ring |
| 5 | Bolt - 6 off | 15 | Lock nut | 25 | Drain plug and level plug |
| 6 | Tab washer - 6 off | 16 | Socket screw - 10 off | 26 | Sealing washer - 3 off |
| 7 | Journal ball bearing - 2 off | 17 | Socket screw | 27 | Breather/detent housing bolt |
| 8 | Journal ball bearing | 18 | Neutral indicator switch | 28 | Breather hose |
| 9 | Dowel - 2 off | 19 | Gasket | 29 | Detent spring |
| 10 | Gasket | 20 | Washer - 2 off | 30 | Plunger |
| | | | | 31 | Cable clip |

6.2a Input shaft bearing retained by plate

6.2b After heating casing, bearings may be displaced

bearing assembly as a unit and position it on the pushrod end.
Grease the assembly and push it home into the end cover. Fit
the clutch operating arm and return spring, using a new split pin
to secure the clevis pin. The clevis pin should be lubricated with
graphite grease.

10 Replace the speedometer driveshaft and housing. Retain the
small end float shim on the shaft with a dab of grease. This will
aid reassembly.

11 The gearbox is now complete with the exception of the
splined boss which is fitted to the input shaft. If the clutch was
dismantled and has yet to be reassembled, the splined boss should
be used to centralise the clutch friction plates prior to the boss
being replaced on the shaft. When refitting the boss, lubricate
the plain collar portion where it passes through the oil seal. Do
not omit to bend up the tab washer to secure the nut, after
tightening.

7.1 Mainshaft and layshaft - general view

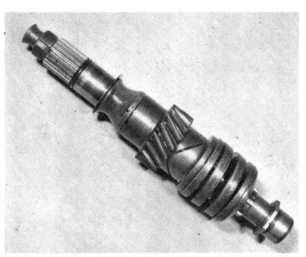

7.4a Place 'O' ring and shim on input shaft

7.4b Lubricate the oil seal lip and ...

7.4c ... insert the shaft into the casing

7.4d Apply grease to hold thrust bearing on layshaft

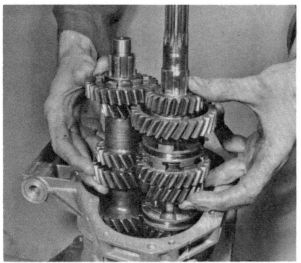
7.4e Insert mainshaft and layshaft simultaneously

7.5a Fit the two mainshaft selector forks and ...

7.5b ... insert the selector drum

7.5c Replace the fork rod and drum endfloat shims

7.5d Refit the layshaft selector fork and dog clutch

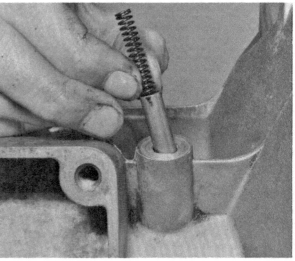
7.6a Insert the detent plunger and spring and ...

7.6b ... secure them with the breather union

7.8 Fit the endcover and a new gasket

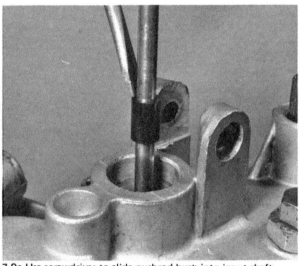

7.9a Use screwdriver to slide pushrod bush into input shaft

7.9b Fit inner thrust piece and thrust bearing

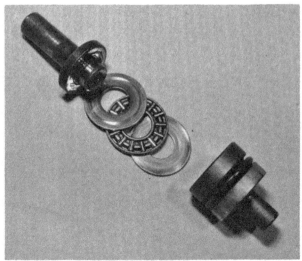

7.9c Pushrod thrust piece components - general view

7.9d Grease completed unit before inserting

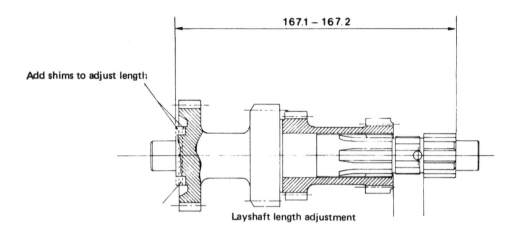

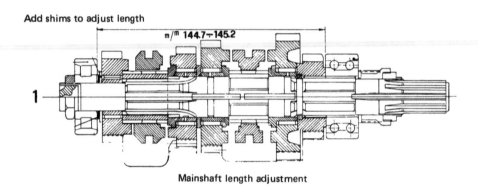

**Fig. 2.3. Layshaft adjusted length and mainshaft adjusted length**

---

### 8  Clutch: removal, examination and replacement, 5-speed gearbox only

1   The clutch components are retained by eight bolts passing through the periphery of the starter ring gear into the flywheel face. Unscrew the eight bolts and lift off the starter ring gear. Remove the following components consecutively; outer friction plate, intermediate plate, inner friction plate, clutch thrust piece, spring back plate and eight clutch springs. Because the complete assembly is under pressure from the clutch springs, the eight bolts must be unscrewed evenly about one turn at a time, so that the spring pressure is released in a controlled and even manner. Note that the spring back plate has an index mark which should be in line with the TDC mark on the flywheel.

Alignment of these two marks ensures the clutch springs will enter the recesses in the back plate inner face.
2   After an extended period of service, the two friction plates, will wear, which may cause slip and in extreme cases, clutch snatch. If slippage is allowed to continue, damage may occur to the spring back plate and starter ring gear faces, due to overheating. Check the width of each friction plate, using a vernier gauge or micrometer. When new, the friction plates measure 8.0 mm (0.3149 in). If either plate has worn by more than 0.5 mm (0.020 in) both plates should be renewed. The friction plates fitted to the 850 Le Mans models are of a heavier material than that utilised for all other models. These plates have a longer expected life and can be fitted as replacements to any 5-speed model.
3   Inspect the intermediate plate for signs of blueing (over-

heating) or scoring on the faces. Using a straight edge, check for warpage of the plate, which will cause clutch drag and noisy gearchanges.

4 The uncompressed or free length of the clutch springs should be checked against that of a new component. If any spring has taken a marked set, the springs should be renewed as a set and not individually.

5 If the faces of the starter ring gear and spring back plate have become scored, these must be renewed. Check the condition of the internal and external teeth on the plates and the teeth on the flywheel and splined gearbox input shaft boss. Excessive wear on these will induce a heavy clutch action and slow disengagement.

6 The clutch release mechanism, which is housed in the gearbox end cover, is unlikely to give trouble unless lubrication failure has caused damage to the thrust bearing. The clutch pushrod should be checked for straightness by rolling it on a flat surface. A bent rod will cause heavy clutch action and must be renewed.

7 Replacement of the clutch is straightforward, the procedure being essentially a reversal of the dismantling operation. Install each of the eight clutch springs in the recesses in the flywheel face. Apply a small amount of heavy graphite grease to

the internal splines in the flywheel. Apply the grease sparingly, to prevent any finding its way onto the clutch plates. Insert the clutch spring plate into position so that the index mark on the plate aligns with the TDC mark on the flywheel. This will aid alignment of the springs with the recesses in the rear of the spring plate.

8 Replace the two friction plates and the intermediate plate, followed by the starter ring gear. Insert and start the eight bolts, but do not tighten them at this stage. Because the two friction plates are fully floating when not under spring pressure, they must be centralised before tightening down the starter ring gear, to enable easy entry of the splined boss of the gearbox input shaft. A special tool is recommended with which to accomplish this task. If this tool is not available, the splined boss may be removed from the gearbox input shaft and used as the centraliser. When using the latter method, first insert the clutch thrust piece to aid centralisation of the boss.

9 With the tool in place, tighten down the eight bolts evenly, about one turn at a time against the spring pressure. The centralising tool should be an easy sliding fit if the plates are positioned correctly. Remove the tool and insert the clutch thrust piece if this has not been done already.

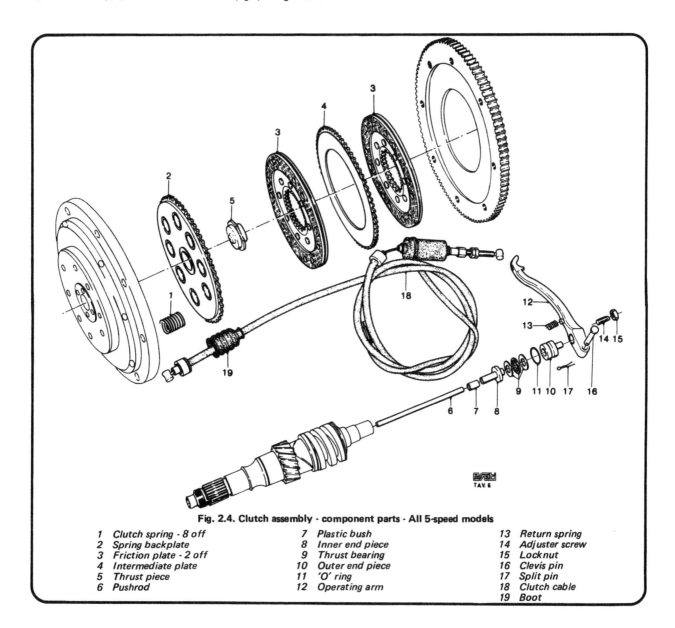

**Fig. 2.4. Clutch assembly - component parts - All 5-speed models**

| | | | | | |
|---|---|---|---|---|---|
| 1 | Clutch spring - 8 off | 7 | Plastic bush | 13 | Return spring |
| 2 | Spring backplate | 8 | Inner end piece | 14 | Adjuster screw |
| 3 | Friction plate - 2 off | 9 | Thrust bearing | 15 | Locknut |
| 4 | Intermediate plate | 10 | Outer end piece | 16 | Clevis pin |
| 5 | Thrust piece | 11 | 'O' ring | 17 | Split pin |
| 6 | Pushrod | 12 | Operating arm | 18 | Clutch cable |
| | | | | 19 | Boot |

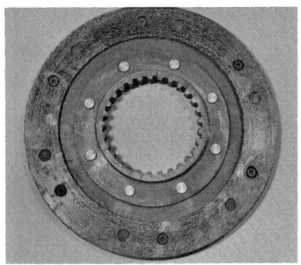

8.2 Check friction plates for spline wear and lining wear

8.7a Position the clutch springs in the flywheel

8.7b Fit the spring backplate so that ...

8.7c ... the dot aligns with the flywheel TDC mark

8.8a Replace friction plates and intermediate plate

8.8b Insert the clutch thrust piece

8.8c Use gearbox input boss to centralise the plates

### 9 Torque converter: principle of operation, V-1000 Convert model only

The torque converter consists of three fans running in an oil filled torroidal or doughnut shaped container bolted to the engine flywheel. The first fan, the turbine, is fixed to the crankshaft, the second fan - the pump - to the gearbox input shaft. As the engine is run at increasingly higher speed, the oil is forced out of the vanes of the turbine into those of the pump, and so drive is transmitted. The third fan, or stator, is free running in one direction and directs the oil across from the pump to the turbine. Except at the 'lock-up' point, when the relative speeds of the drive and driven components are almost the same, large amounts of heat are produced in the torque converter due to frictional losses. To prevent the fluid boiling, a cooling system is incorporated. It comprises an externally - mounted oil cooler, fed by an engine driven pump, through which the fluid from the converter is passed before being returned to the reservoir which feeds the torque converter.

### 10 Torque converter: removal, examination and replacement, V-1000 convert model only

1   The torque converter will only require removal if the engine flywheel is to be removed or if the converter itself requires renewal. Examination of the torque converter may be made with the unit in situ.
2   Check that the outer surface of the central boss is not scored and that a radial groove has not been worn in the metal by the oil seal lip. An imperfect boss will promote seepage of the hydraulic fluid.
3   Inspect the two internal needle roller bearings for damaged rollers. Slide the clutch input shaft into position in the converter and check for up and down play. Wear of the bearings is unlikely to occur until a substantial mileage has been covered. Check also that the tracks on the clutch input shaft where the bearings resolve are not worn or scuffed.
4   The torque converter is a sealed unit and therefore if any component fails the complete assembly must be renewed.
5   The torque converter is retained by four bolts which pass through the starter ring gear and into the periphery of the flywheel. Remove the bolts after bending down the ears of the locking plates.
6   It is recommended that on removal of the gearbox a cover be placed over the aperture in the torque converter to prevent the ingress of foreign matter.

7   To replace the torque converter on the engine, position the starter ring gear and the torque converter on the flywheel and insert and tighten the bolt lightly. It is essential that the boss of the torque converter into which the gearbox input shaft fits, runs absolutely concentrically. If this precaution is not taken difficulty may be encountered when trying to fit the gearbox to the engine. Additionally, oil leakage will occur at the oil seal. To check for correct rotation a dial gauge or index gauge should be mounted on a bracket retained on one of the gearbox mounting studs. If eccentricity is apparent, remove the four retaining bolts and move the torque converter one quarter of a turn until the bolt holes again line up. Refit the bolts and check again. The maximum permissible run out measured on the outer surface of the boss is 0.05 - 0.06 mm (0.0019 - 0.0023 in).
8   When the torque converter is running true, tighten the bolts and secure them by means of the locking tabs.

10.4 Torque converter retained by four bolts secured by lock-plates

### 11 Clutch: removal, examination and reassembly, V-1000 Convert model only

1   To gain access to the clutch the gearbox bell housing must be detached from the gearbox together with the clutch input shaft and the torque converter static shaft. Detach the pipe from the base of the bellhousing by unscrewing the banjo bolt from the union. Loosen and remove the six bolts which pass from the bellhousing into the gearbox. Using a rawhide mallet separate the two components which are located by two dowels in the mating surfaces.
2   Rotate the clutch centre boss, and by passing a spanner through one of the holes in the boss remove each of the six flange plate retaining bolts. Pull the centre boss from position complete with the integral shaft. The ball bearing and flange may be removed from the shaft after detaching the 'O' ring and circlip.
3   The clutch plates are secured in the clutch outer drum by a large internal circlip. Because the clutch plates are under pressure from the clutch springs the circlip cannot be removed until the pressure has been released. If an attempt is made to prise the circlip from position without releasing pressure, when freed the clutch plates will spring out causing damage to the components and also the operator.
4   To release the pressure on the clutch plates, release the lock-nut on the end of the clutch operating pull rod and tighten the adjuster down. In doing this the spring back plate will be drawn against the clutch springs. When it can be seen that the clutch plates are no longer under pressure, prise the circlip from place using a screwdriver, and remove the clutch plates one at a time, noting their relative positions for easy reassembly. Slacken off

the adjuster and remove it and the locknut from the clutch pull-rod. Withdraw the rod and the ball bearing from the spring back plate.

5    Lift the spring back plate from the clutch outer drum and remove the clutch springs. Note the shims, one of which lies between each spring and the back plate recess. The clutch outer drum is retained on the splined gearbox input shaft by a shouldered nut. The nut is secured by a portion of the shoulder being bent into the axial groove in the shaft. To prevent the input shaft rotating when loosening the nut, refit temporarily the final driveshaft, and place the shaft in the soft jaws of a vice. Care should be taken not to damage the shaft by over-tightening. Place the gearbox in high gear and loosen the nut after bending the shoulder away from the shaft groove. After removal of the nut the clutch outer drum may be pulled off the shaft.

6    Commence clutch inspection by checking the friction plates. After an extended period of service, the clutch linings will wear and promote clutch slip. The limit of wear measured across each plate is 2.65 mm (0.104 in). When the overall width of the linings reach this level, the clutch plates must be replaced as a complete set.

7    The plain clutch plates should not show any evidence of excess heating (blueing) and should not be more than 0.05 mm (0.002 in) out of true.

8    The clutch springs should have a free (uncompressed) length of 28.00 mm (1.102 in). If the springs have taken a set of 1mm

(0.040 in) or more, the complete set must be renewed.

9    Check the condition of the slots in the outer surface of the clutch centre and the inner surfaces of the outer drum. In an extreme case, clutch chatter may have caused the tongues of the inserted plates to make indentions in the slots of the outer drum, or the tongues of the plain plates to indent the slots of the clutch centre. These indentations will trap the clutch plates as they are freed and impair clutch action. If the damage is only slight the indentations can be removed by careful work with a file and the burrs removed from the tongues of the clutch plates in similar fashion. More extensive damage will necessitate renewal of the parts concerned.

10   Check the clutch disengagement bearing for wear or rough-ness of rotation. Inspect the pullrod anchor bush, which is a drive fit in the bearing, for looseness and for wear on the surface which engages with the pullrod head. Check also the pullrod for similar wear and for straightness.

11   The clutch may be reassembled by reversing the dismantling procedure. When inserting the clutch plates commence by refitting a friction plate followed by a plain plate and so on. Fit the double thickness plain plate last and then replace the circlip. Ensure that the circlip is fully home in its seating groove before releasing the pullrod. When the clutch plates are still free, and the springs are still compressed align the internal tongues of the plain plates, using a straight edge, so that the clutch boss may enter the clutch easily on reassembly.

11.2 Clutch boss retainer plate held by six obscured bolts

11.4a Screw down adjusting nut before releasing circlip

11.4b Remove all plates noting original sequence

11.4c Remove pullrod from backplate and ...

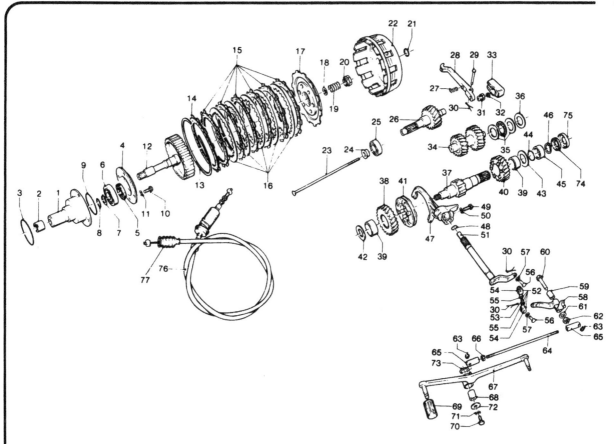

**Fig. 2.5. Gearbox and clutch components - V-1000 model**

| | |
|---|---|
| 1 | Torque converter static shaft |
| 2 | Bush |
| 3 | 'O' ring |
| 4 | Flange plate |
| 5 | Oil seal |
| 6 | Journal ball bearing |
| 7 | Circlip |
| 8 | 'O' ring |
| 9 | 'O' ring |
| 10 | Bolt - 5 off |
| 11 | Star washer - 5 off |
| 12 | Clutch input shaft |
| 13 | Circlip |
| 14 | Outer plate |
| 15 | Friction plate - 6 off |
| 16 | Plain plate - 5 off |
| 17 | Pressure plate |
| 18 | Shim - 6 off |
| 19 | Spring - 6 off |
| 20 | Nut |
| 21 | 'O' ring |
| 22 | Clutch outer drum |
| 23 | Pull rod |
| 24 | Pull piece |
| 25 | Journal ball bearing |
| 26 | Gearbox input shaft |
| 27 | Return spring |
| 28 | Clutch operating arm |
| 29 | Pivot pin |
| 30 | Split pin |
| 31 | Adjuster nut |
| 32 | Locknut |
| 33 | Rubber boot |
| 34 | Layshaft |
| 35 | Thrust bearing |
| 36 | Thrust washer |
| 37 | Mainshaft |
| 38 | Low speed gear pinion |
| 39 | Bush |
| 40 | High speed gear pinion |
| 41 | Sliding dog clutch |
| 42 | Shim |
| 43 | Shim |
| 44 | 'O' ring |
| 45 | Spacer |
| 46 | Circlip |
| 47 | Gear selector fork |
| 48 | 'O' ring |
| 49 | Pinch bolt |
| 50 | Star washer |
| 51 | Selector shaft |
| 52 | Vertical link rod - complete |
| 53 | Link rod |
| 54 | Clevis fork - 2 off |
| 55 | Locknut - 2 off |
| 56 | Clevis pin - 2 off |
| 57 | Washer - 2 off |
| 58 | Intermediate arm |
| 59 | Bush |
| 60 | Bolt |
| 61 | Washer |
| 62 | Nut |
| 63 | Circlip |
| 64 | Horizontal link rod |
| 65 | Joint bar - 2 off |
| 66 | Lock nut |
| 67 | Gearchange lever |
| 68 | Bush |
| 69 | Rubber |
| 70 | Bolt |
| 71 | Star washer |
| 72 | Washer |
| 73 | Shim |
| 74 | Rubber shroud |
| 75 | Shroud |
| 76 | Clutch cable |
| 77 | Rubber boot |

11.5a ... lift out bearing and spring backplate

11.5b Note the shims between plate and springs

11.5c Clutch drum is retained by shouldered nut

11.10 Pullrod bush is a drive fit in bearing

## 12 Gearbox: dismantling, examination and reassembly, V-1000 Convert model only

1    After removal of the clutch, the gearbox is ready for dismantling. Remove the clutch operating arm by detaching the clevis pin which is secured by a split pin. Unscrew the upper chrome bolt from the gearbox end cover and invert the gearbox so that the detent spring and plunger fall out. Prise the shroud and rubber spacer from the casing boss concentric with the output shaft.

2    Remove the gearbox end cover screws and using a rawhide mallet, separate the cover from the main casing. Note and remove the thrust washers, shims and thrust bearing on the end of the layshaft. Drift the input shaft from position in the bearing using a rawhide mallet. Remove the circlip and spacer from the output shaft and draw off the bearing; this will allow the high speed gear pinion and spacer to be withdrawn.

3    Pull the layshaft from position. Loosen the selector fork pinch bolt and pull the selector shaft from the fork. Grasp the end of the output shaft and pull the unit from position complete with the selector fork. Separate the selector fork and sliding dog from the upper end of the output shaft, and the shim and low-speed gear pinion from the lower end.

4    If required, the speedometer driveshaft may be removed by unscrewing the housing from the outside of the gearbox casing.

5    Examine the gear pinions to check that there are no broken or chipped teeth, and that the selector dogs on the gear faces are not rounded. Repairs to damage of this nature are not practicable, and as such the components in question must be renewed. All the free-running gear pinions are supported on plain bushes. If wear is evident in the form of scoring or scuffing, the bushes must be renewed. Check also that the shafts and pinions with which the bushes work are not damaged.

6    Inspect the selector fork for wear of the fingers, and for wear of the channel in the sliding dog clutch with which the fingers locate. Excessive clearance between the two components will induce less positive gear selection. Check also the projection on the fork which locates with the spring loaded detent plunger. The selector fork, together with the sliding dog clutch, was modified from gearbox No. G01001 onwards so that the finger ends locating with the sliding dog clutch can be renewed independently from the main fork. The modified components may be refitted in place of earlier components.

7    Refer to Section 7 of this Chapter for comments on bearing and oil seal inspection and removal.

8    Reassemble the gearbox by reversing the dismantling procedure. The shim fitted on the output shaft between the low

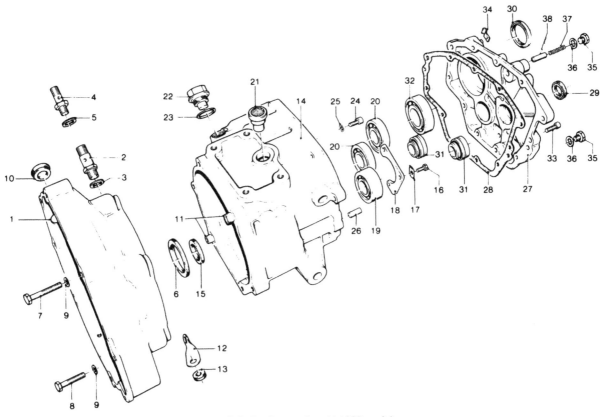

**Fig. 2.6. Gearbox casing - V-1000 models**

| | | | | | | | |
|---|---|---|---|---|---|---|---|
| 1 | Bellhousing | 11 | Dowel - 2 off | 21 | Breather plug | 31 | Needle roller bearing - 2 off |
| 2 | Fluid union | 12 | Cable guide | 22 | Oil filler cap | 32 | Journal ball bearing |
| 3 | Sealing washer | 13 | Grommet | 23 | Sealing washer | 33 | Socket screw - 11 off |
| 4 | Fluid union | 14 | Clutch/gearbox housing | 24 | Socket screw | 34 | Lead clip |
| 5 | Sealing washer | 15 | Oil seal | 25 | Star washer | 35 | Plug - 2 off |
| 6 | Oil seal | 16 | Bolt - 3 off | 26 | Dowel - 2 off | 36 | Sealing ring - 2 off |
| 7 | Bolt - 2 off | 17 | Tab washer - 3 off | 27 | End cover | 37 | Detent spring |
| 8 | Bolt - 4 off | 18 | Bearing retainer plate | 28 | Gasket | 38 | Detent plunger |
| 9 | Plain washer - 6 off | 19 | Journal ball bearing | 29 | Oil seal | | |
| 10 | Inspection plug | 20 | Journal ball bearing | 30 | Oil seal | | |

speed gear and the gearbox wall should be replaced with the chamfered edge facing the gear. The layshaft endfloat when in place should be 0.15 - 0.20 mm (0.005 - 0.007 in). Remove or add shims between the gearbox end cover and the needle thrust bearing outer washer to arrive at the correct clearance. The selector rod should be fitted into the selector fork so that the arm is in exact alignment with the plunger locating projection on the fork.

---

### 13 Gearbox: refitting to engine and adjusting the clutch, all models

1 Provided that the clutch plates have been aligned correctly, refitting of the gearbox onto the engine is straightforward. On 5-speed models if difficulty is encountered when inserting the splined boss into the clutch centre, a long handled screwdriver may be inserted between the two mating faces with which to rotate the flywheel.

2 On automatic models, the gearbox must be refitted to the bellhousing and secured by means of the six bolts. Before replacing the completed assembly, lubricate the outer surface of the torque converter central boss so that it enters the oil seal in the bellhousing easily. Ensure that the bronze bush on

the clutch shaft is fitted with the cut-aways outwards.

3 The adjustment of the clutch operating mechanism on all models should be made with the engine/gearbox unit in the frame and the clutch cable connected. On 5-speed models, loosen the locknut on the clutch operating arm and by means of the adjuster screw, set the arm so that at the point of clutch disengagement when the clutch lever is operated, the arm is at about 90° to the pushrod. This will give the greatest leverage and hence the most sensitive control of the clutch. Tighten the locknut and then adjust the cable at the handlebar lever adjuster so that there is 4 mm (1/8 in) movement at the ball end of the lever before clutch disengagement commences.

4 On V-1000 models, loosen the locknut on the adjuster and screw the adjuster in or out until the distance between the inner face (rear) of the cable abutment lug on the gearbox and the curved portion of the operating arm against which the cable nipple seats is as follows:

| | |
|---|---|
| 30 mm | On a used machine |
| 33 mm | On a new machine or when new clutch plates have been installed |

Tighten the locknut and then adjust the cable as described for 5-speed machines.

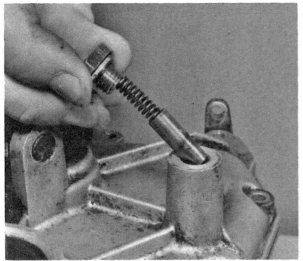

12.1a Unscrew plug to remove detent assembly

12.1b Gearbox components - general view

12.2 Note thrust bearing on layshaft

12.8a Input shaft bearing secured by retainer plate

12.8b Pawl tooth on selector should line up with arm

13.3 Cut-aways on bush must face outwards

13.4 Adjust clutch arm by means of adjuster nut

**14 Torque converter oil pump: removal, examination and replacement, V-1000 Convert model only**

1  The torque converter oil pump is housed within a chamber in the valve timing cover and is driven directly from the forward end of the camshaft via an hexagonal drive piece. To gain access to the oil pump, the timing cover must be removed. Ideally the engine/gearbox unit should be separated from the frame for this to be carried out as described in Chapter 1, Section 4. It is possible however, if a little ingenuity is exercised, to remove the timing cover with the engine/gearbox unit still in place. To do this, the engine must be supported on blocks and the front engine mounting bolt removed. In addition, the torque converter fluid cooler must be detached to aid access. See Chapter 3, Fig. 3.5 for diagram of oil pump.
2  Before disconnecting any feed or return pipes in the fluid circuit, the system must be drained. Remove the left-hand frame cover to gain access to the fluid reservoir. Unscrew the lower banjo bolt and allow the fluid to drain from the reservoir. Detach the upper hose from the cooler unit to allow all the fluid to be drained through the cooler feed pipe, which has already been disconnected.
3  Drain the engine oil into a suitable container and then remove the alternator, as described in Chapter 1, Section 8. The timing cover is retained by fourteen socket screws. After removal of the screws use a rawhide mallet to separate the cover from the engine. As the cover is removed, the hexagonal drive piece will remain in either the camshaft or the oil pump.
4  Before attending to the oil pump, unscrew the large domed nut from the outside of the timing cover and remove the detent spring and pressure release valve ball.
5  Unscrew the three bolts which secure the pump cover to the timing cover casing. Some models incorporate a one-piece lock plate to secure the three bolts. In this case the ears on the plate must be bent down before the nuts are loosened. Lift the outer cover from position complete with the driveshaft and the drive pin. The cover may be a tight fit in the casing. Unfortunately, no method is provided by which a tight fitting cover may be removed with ease. Very carefully use a sharp tipped screwdriver between the two mating surfaces to help displace the cover. This may damage the material, although unsightly, slight damage in this area is not critical because the faces do not comprise a sealing joint. After removal of the cover, detach the large 'O' ring from the cover spigot and displace the drive pin and the driveshaft.

6  Lift the inner rotor from place followed by the outer rotor. Clean all the components in petrol and allow them to dry before inspection.
7  Examine the two rotors for scoring, caused by swarf finding its way into the oil pump. Damage of this nature is unlikely unless contamination of the circuit has occurred during overhaul. Check also that the bore in the timing cover is not scored. The oil pump outer rotor runs directly in the timing cover, therefore, if damage occurs to the housing, the complete cover must be renewed. Check the diametrical clearances between the outer rotor and pump housing and between the outer and inner rotors, with a feeler gauge. The specifications are given at the beginning of the Chapter.
8  Check the driveshaft outer surface for scoring and ensure that an annular groove has not been worn by the oil seal lip. The driveshaft runs directly in the pump cover. If the clearance exceeds 0.145 mm( 0.006 in), the shaft or pump cover should be renewed.
9  If damage to the oil seal occurs, hydraulic fluid will find its way into the engine oil causing a drop in the reservoir level and an undesirable contamination of the engine lubricant. The seal can be prised from position after removing the internal circlip with a suitable pair of pliers. Check carefully that the small oil feed hole in the pump cover has not become obstructed. Failure of oil to reach the driveshaft will cause seizure of the pump shaft and consequent temporary failure of the torque converter, due to a lack of fluid. The oil pump drive piece is made from a soft steel, the edges of which will round off if seizure occurs, and so prevent actual breakage of the drivepiece or driveshaft.
10  Reassemble the oil pump by reversing the dismantling procedure. Lubricate all the components with hydraulic fluid during reassembly. The two rotors must be fitted so that the punch-marked face of each is away from the timing cover. Ensure that the sealing 'O' ring is in good condition, before replacement.

**15 Torque converter circuit pipes and seals: examination, V-1000 Convert model only**

1  Inspect the feed and return pipes periodically for signs of perishing or damage caused by trapping or bad routing. When refitting pipes, ensure that they are positioned correctly away from hot components and that there are no kinks. Use new sealing washers at the banjo unions, whenever there is any doubt about their condition. Scored or compressed washers will promote leakage and encourage overtightening of the banjo bolts, which shear easily.

14.3 Hexagonal converter oil pump drive piece

14.4 Remove the pressure release valve from timing cover

14.5a Pump cover is retained by three bolts

14.5b Withdraw housing - note the 'O' ring

14.7a Using a feeler gauge check inner and ...

14.7b ... outer rotor for wear

14.9 Oil seal is retained by a circlip

14.10 Fit rotors with both punch marks facing outwards

**Fig. 2.7. Torque converter coolant system**

| | | | |
|---|---|---|---|
| 1 | Fluid reservoir | 15 | Sealing washer - 3 off |
| 2 | Filler cap/dipstick | 16 | Return pipe |
| 3 | 'O' ring | 17 | Banjo bolt - 4 off |
| 4 | Bolt | 18 | Sealing washer - 11 off |
| 5 | Plain washer | 19 | Domed nut - 2 off |
| 6 | Rubber sleeve - 2 off | 20 | Radiator return pipe |
| 7 | Bolt | 21 | Radiator feed pipe |
| 8 | Plain washer | 22 | Converter feed pipe |
| 9 | Warning sticker | 23 | Breather pipe |
| 10 | Rubber cushion | 24 | Strap |
| 11 | Banjo bolt with filter | 25 | Oil cooler |
| 12 | Sealing washer - 2 off | 26 | Bolt - 2 off |
| 13 | Pump feed pipe | 27 | Plain washer - 2 off |
| 14 | Banjo bolt | 28 | Rubber buffer - 4 off |

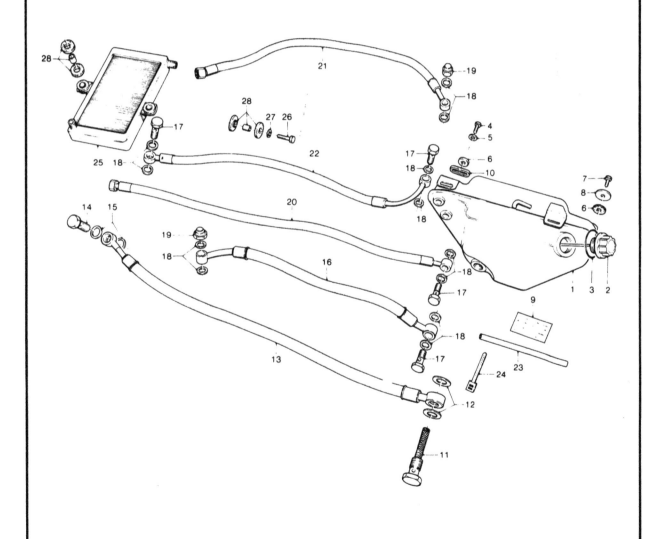

### 16 Torque converter reservoir: location and filter cleaning, V-1000 Convert model only

1   The torque converter fluid reservoir is retained on the left-hand side of the machine under the frame side-cover. Access may be gained by pulling the cover from position. The tank filler cap incorporates a dipstick to facilitate level checking.
2   The oil feed line banjo bolt (the lower bolt) incorporates a cylindrical gauze filter, which should be removed and cleaned in petrol whenever the fluid is renewed.

### 17 Torque converter oil cooler: removal and examination, V-1000 Convert model only

1   The torque converter hydraulic fluid cooler is similar in construction to the water radiator utilised on most motor cars.

It is mounted on the forward frame down tubes to take advantage of the good air flow.
2   To remove the oil cooler, first drain the fluid by disconnecting the right-hand pipe followed by the left-hand pipe. Both pipes are retained by concentric olive type unions. Tie the pipes up above the level of the reservoir, to prevent all the fluid draining from the system. The cooler is retained on the frame by two bolts passing through rubber bushes.
3   Repair of a leaking cooler is impracticable and as such a damaged unit must be renewed. It is not possible to introduce a sealant into the system because of contamination.
4   After a period of service, the external cooling fins will become blocked by wind-blown matter causing a severe reduction in cooling efficiency. Frequent overheating of the fluid, evident by boiling noises eminating from the reservoir, can often be traced to this fault. Clean the outside of the cooler in a grease solvent, and after rinsing, dry, using a high pressure air hose. **Do not** apply the pressure hose to either the inlet or outlet as the pressure will damage the matrix.

16.2 Lower pipe banjo bolt incorporates filter screen

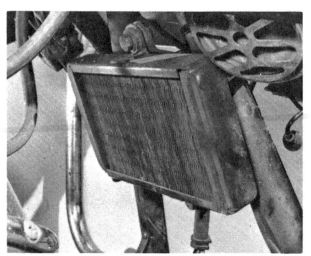

17.2 Oil cooler mounted on rubber bushes

### 18 Fault diagnosis - gearbox

| Symptom | Cause | Remedy |
| --- | --- | --- |
| Difficulty in engaging gears | Gear selectors not indexed correctly | Check alignment. |
|  | Selector forks bent or badly worn | Renew. |
| Machine jumps out of gear | Camplate plunger sticking | Remove and free. |
|  | Worn dogs or dogs on gear pinions | Renew defective pinions or dogs. |
| Gear change lever does not return to original position | Broken return spring | Renew spring. |

**19 Fault diagnosis: clutch**

| Symptom | Cause | Remedy |
|---|---|---|
| Clutch slip | Oil on plates | Remove and clean. |
| | | Renew faulty seals. |
| | Lack of free play on operating rod | Readjust control arm or cable. |
| | Worn plates | Renew. |
| | Weak springs | Renew. |
| Operating action stiff | Bent pushrod | Renew |
| | Dry pushrod | Grease. |
| | Dry plate splines (not V-1000) | Grease. |
| | Damaged, trapped or frayed control cables | Check cable and renew if required. Ensure cable is lubricated and has no sharp bends. |

**20 Fault diagnosis: torque converter - V-1000 Convert model only**

| Symptom | Cause | Remedy |
|---|---|---|
| No drive to rear wheel | Fluid leakage | Check level in reservoir and inspect for leakage. |
| | Sheared drive piece | Remove timing cover and inspect. |
| Fluid boils consistently | Blocked oil cooler fins | Clean. |

# Chapter 3 Fuel system and lubrication

**Contents**

**Specifications**

## V-1000 I-Convert, 850T, T-3 and 750 S and S3 models

### Carburettor

| | |
|---|---|
| Make ... ... ... ... ... ... ... ... ... | Dell 'Orto |
| Type ... ... ... ... ... ... ... ... ... | VHB 30CD (right-hand)  VHB 30CS (left-hand) |
| Atomiser ... ... ... ... ... ... ... ... ... | 265 |
| Idling jet ... ... ... ... ... ... ... ... ... | 50 |
| Starting jet ... ... ... ... ... ... ... ... | 80 |
| Needle ... ... ... ... ... ... ... ... ... | V9, 2nd notch (V-1000 I-Convert model, U-9) |
| Choke ... ... ... ... ... ... ... ... ... | 40 |

| | 750 S, S3 & 850T | 850T3 | V-1000 |
|---|---|---|---|
| Main jet ... ... ... ... ... ... ... ... ... | 142 | 120 | 130 |

| | |
|---|---|
| Mixture screw; no. of turns out: | |
| 750S and 850T ... ... ... ... ... ... ... | 2-2½ (LH), 2¼-2¾ (RH) |
| 750 S3 ... ... ... ... ... ... ... ... | 2-2½ (LH), 2-2¾ (RH) |
| 850T3 and V-1000 ... ... ... ... ... ... ... | 1½ |
| Air filter type ... ... ... ... ... ... ... ... | Oil impregnated plastic foam |

## Le Mans models

### Carburettor

| | |
|---|---|
| Make ... ... ... ... ... ... ... ... ... | Dell 'Orto |
| Type ... ... ... ... ... ... ... ... ... | PHF 36 B (D) right-hand, PHF 36 B (S) left-hand |
| Atomiser ... ... ... ... ... ... ... ... ... | 265 AB |
| Main jet ... ... ... ... ... ... ... ... ... | 135 |
| Pilot jet ... ... ... ... ... ... ... ... ... | 60 |
| Starter jet ... ... ... ... ... ... ... ... | 70 |
| Accelerator pump jet ... ... ... ... ... ... ... | 38 |
| Mixture screw: no. of turns out ... ... ... ... ... | 1½ |
| Needle ... ... ... ... ... ... ... ... ... | K5 (2nd notch) |

### Lubrication system

| | |
|---|---|
| Type ... ... ... ... ... ... ... ... ... | Wet sump, high pressure |
| Filter ... ... ... ... ... ... ... ... ... | Gauze screen, sealed cartridge (except 750S and some T models) |
| Oil capacity: | |
| 750 S, S3 and 850 T ... ... ... ... ... ... | 3.5 lit (3.70 US qt/6.16 Imp pt) |
| All other models ... ... ... ... ... ... ... | 3.0 lit (3.17 US qt/5.28 Imp pt) |
| Oil pump type ... ... ... ... ... ... ... ... | Helical gear |

| | | | | | | | | |
|---|---|---|---|---|---|---|---|---|
| Oil pressure | ... | ... | ... | ... | ... | ... | ... | ... | 55-60 psi (3.8-4.2 kg cm²) |
| Pump body depth | ... | ... | ... | ... | ... | ... | ... | 14.032-14.075 mm (0.5524-0.5541 in) |
| Pump gear width | ... | ... | ... | ... | ... | ... | ... | 13.973-14.000 mm (0.5501-0.5511 in) |
| Gear diameter | ... | ... | ... | ... | ... | ... | ... | 26.250-26.290 mm (1.0334-1.0350 in) |
| Pump body internal diameter | ... | ... | ... | ... | ... | ... | 26.340-26.390 mm (1.0370-1.0389 in) |
| Driven gearshaft diameter | ... | ... | ... | ... | ... | ... | 9.985-10.000 mm (0.3931-0.3937 in) |
| Shaft recess in body | ... | ... | ... | ... | ... | ... | ... | 10.013-10.035 mm (0.3942-0.3950 in) |
| Shaft/recess clearance | .... | ... | ... | ... | ... | ... | ... | 0.013-0.050 mm (0.0005-0.0019 in) |

## 1  General description

The fuel system comprises a petrol tank fitted with two taps from which the two carburettors are fed by gravity via interconnected petrol pipes. On 750S and V-1000 models the left-hand tap is of the electrovalve type and is operated by an integral solenoid when the ignition is switched on. All other taps are manually operated and combine a reserve position to provide a small amount of fuel after the main supply has run dry. A float-operated switch, mounted in the under side of the tank and interconnected with a warning light on the handlebar console, is fitted to V-1000 models, to indicate when the reserve fuel level is reached. A cylindrical gauze filter is fitted to each tap within the petrol tank, and a secondary filter is incorporated at each carburettor union.

The two carburettors breathe through a replaceable dry paper cartridge air filter on all models but the 850 Le Mans model, which has a gauze-screened bell mouth fitted to each carburettor. A collector box is incorporated in the air filter system, which allows expelled oil vapour from the crankcase and rocker boxes to be passed through the cylinders where it is burned during the combustion stage and exhausted through the exhaust system.

Dell 'Orto concentric, slide-type carburettors are fitted to all models. The carburettor throttle valve on each unit is controlled by a separate cable connected to a shared twist grip throttle control on the handlebars. The carburettor fitted to the 850 Le Mans models differs from all others in having a throttle slide controlled accelerator pump to improve acceleration. Each carburettor is fitted with a choke assembly, operated either by a carburettor mounted lever or remotely by a lever mounted on the left-hand rocker cover and connected by control cables.

Engine lubrication is provided by oil contained in the sump and circulated through a gear pump, driven by the timing chain, at the front of the engine. All models are fitted with a gauze oil strainer through which the sump oil is passed before reaching the pump. In addition, all models but the 750S and some 850T models, have a replaceable cartridge-type main feed filter through which the oil is forced before being transmitted to the working parts of the engine. Both gauze screen filter and cartridge filter are contained within the sump.

## 2  Petrol taps: removal, cleaning and replacement

1  Two petrol taps are fitted to all models, of which the left-hand tap on V-1000 Convert and 750S models is electronically operated by a solenoid on the tap body.
2  Before the taps can be unscrewed, the contents of the tank must be drained into a suitable container. Release the spring clips or screw clips from the hoses and pull them from the tap unions. Operate the electrovalve tap by turning on the ignition.
3  Remove the taps by unscrewing the gland nuts. The two wires fitted to the solenoid on electrovalve units are a push fit on the terminals. Each tap is fitted with a gauze mesh filter which should be cleaned in petrol at regular intervals as specified under the routine maintenance heading.
4  If the electrovalve tap malfunctions, it should be replaced by a new unit as repair is impracticable. Slight leakage of the manual tap can often be rectified by tightening the lever securing nut slightly, which may vibrate loose. The taps may be dismantled when in position on the petrol tank, after draining the fuel. Unscrew the lever retaining nut and withdraw the lever, complete with the nut and backing spring. The Neoprene seal can be hooked

out with a small screwdriver. If leakage of the tap occurs, the seal must be renewed.
5  When reassembling the tap, ensure that the holes in the seal are correctly aligned. When screwing in the centre nut take care that it does not become cross-threaded. This is easy to do since the nut is spring loaded and must therefore be pushed inwards as well as turned.

## 3  Carburettors: removal and replacement

1  Before removing the carburettors it is necessary to detach the petrol feed pipes with the unions at the carburettors. Unscrew the slotted bolt which passes through each union and displace the circular gauze filter screen.
2  Unscrew the two carburettor top retaining screws from each instrument and withdraw the tops, complete with throttle slide. Remove the throttle slide return spring followed by the throttle slide and carburettor top from each cable. On Le Mans models this is not necessary as the return springs are securely retained and will not spring from position inadvertently. Where a remote choke control is used, unscrew the choke caps and pull the assemblies from position.
3  On all but Le Mans model, each carburettor is retained on the inlet stub by a screw clamp. Because of the design of the air filter box rubber duct, it is simpler to detach each inlet stub from the cylinder head by removing the three socket screws and then separating the carburettor from the stub. On Le Mans models, each carburettor is retained on the inlet stub by a heavy rubber hose which is secured by two screw clips. The rubber mounting helps isolate the carburettors from vibration.
4  Note the difference between the left and right-hand carburettors and lay them aside carefully.
5  Replace in the reverse order to removal. Insert each throttle slide carefully, entering the needle into the jet. Do not oil the slides. Make sure the lug on the mixing chamber top engages with the notch in the carburettor body.
6  Make sure that all the hose clips are tight, making airtight joints. If this is not so, the mixture will be weakened with possible damage to the pistons. Check that the controls work smoothly.

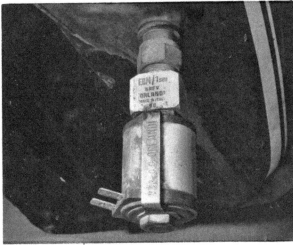

2.1 'Electovalve' ignition operated petrol tap

2.3a Petrol pipes held by screw or spring clips

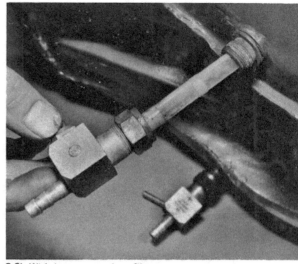

2.3b Withdraw tap to clean filter

3.1 Detach the petrol pipes and unions

3.2a Remove carburettor caps and ...

3.2b ... Disconnect the throttle slides

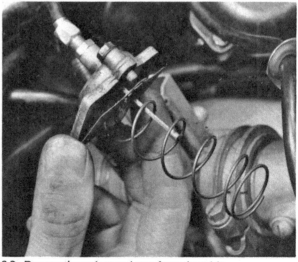

3.2c Remove the springs and caps from the cable

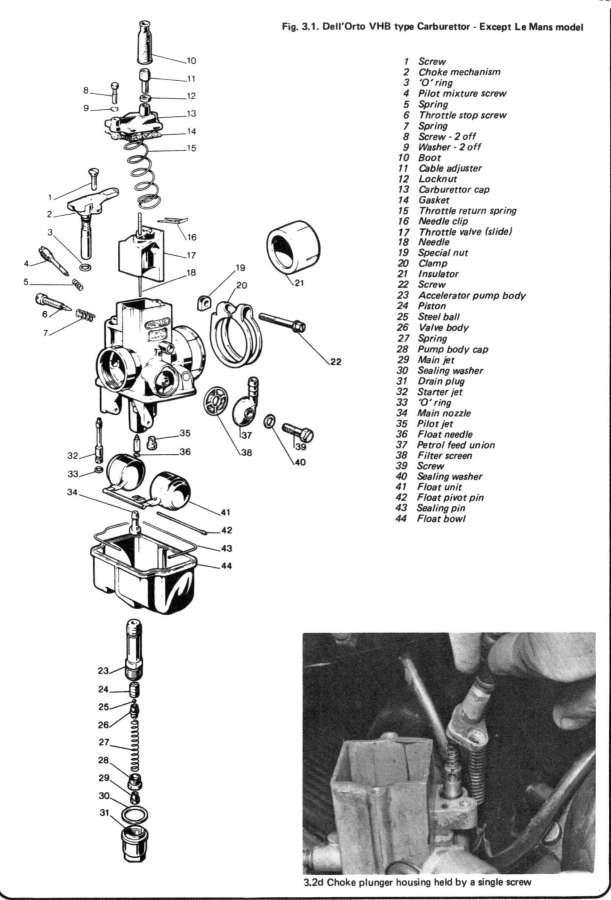

Fig. 3.1. Dell'Orto VHB type Carburettor - Except Le Mans model

1  Screw
2  Choke mechanism
3  'O' ring
4  Pilot mixture screw
5  Spring
6  Throttle stop screw
7  Spring
8  Screw - 2 off
9  Washer - 2 off
10  Boot
11  Cable adjuster
12  Locknut
13  Carburettor cap
14  Gasket
15  Throttle return spring
16  Needle clip
17  Throttle valve (slide)
18  Needle
19  Special nut
20  Clamp
21  Insulator
22  Screw
23  Accelerator pump body
24  Piston
25  Steel ball
26  Valve body
27  Spring
28  Pump body cap
29  Main jet
30  Sealing washer
31  Drain plug
32  Starter jet
33  'O' ring
34  Main nozzle
35  Pilot jet
36  Float needle
37  Petrol feed union
38  Filter screen
39  Screw
40  Sealing washer
41  Float unit
42  Float pivot pin
43  Sealing pin
44  Float bowl

3.2d Choke plunger housing held by a single screw

3.2e Cable operated choke control lever

3.3 Detach the carburettor complete with inlet stub

## 4 Carburettors: dismantling and cleaning.
## Model VHB carburettors only

1   Remove each carburettor as described in the previous Section. Dismantle them separately, so that the parts do not become interchanged. Note that the carburettor marked CD is the right-hand instrument and the one marked CS is the left-hand instrument.
2   Invert the carburettor and unscrew the large nut from the centre of the float chamber. Lift the float chamber bowl from place, noting the sealing O ring. Using a pair of snipe nosed pliers, withdraw the float assembly pivot pin. Lift the floats from position together with the float needle. Displace the needle and store it safely as it is easily lost.
3   From the centre of the mixing chamber base unscrew the main jet and then remove the accelerator pump block as a unit, by fitting a spanner to the two flats on the main body. Separate the accelerator pump parts from the block after removing the end cap which held the main jet. The components include a return spring and a piston containing a non-return ball valve. The main nozzle which protrudes into the venturi of the carburettor, can be pushed out from the venturi side.
4   Unscrew the pilot jet from the boss adjacent to the accelerator pump holder. Then unscrew the choke (starter) jet complete with sealing ring from the turret to one side of the base.
5   The pilot adjuster screw should be unscrewed from the outside of the carburettor body. The choke control lever and plunger assembly (where utilised) may be removed after unscrewing the single screw which passes through the housing cap.
6   Clean all the passageways in the main body and the jets, using a compressed air jet. Do not use stiff wire or other pointed instruments for clearing obstructed jets or passageways. It is only too easy to enlarge the precision drilled orifices and cause carburation changes which will prove very difficult to rectify.
7   Persistent flooding is often caused by a leaking float, which will cause the petrol level to rise, or by dirt on the float needle or need valve seating. A worn needle may also cause flooding due to bad sealing, or may, in unusual circumstances, stick shut, causing a lack of fuel.
8   Before reassembling the carburettor in the reverse order to that given for dismantling, make sure all the component parts are clean. Check that the needle is not bent, by rolling it on a sheet of plain glass. Examine the throttle slide; signs of wear will be evident on the polished outer surface.
9   When replacing the throttle valves, make sure the slot in the base of each valve registers with the projection inside the mixing chamber, so that the valve will seat correctly. It is also important

to check that the needle suspended from each throttle valve has entered the needle valve, otherwise there is risk of damaging both the needle and the jet.

## 5 Carburettors: dismantling and cleaning.
## Model PHF carburettors only

1   To remove the carburettors, see Section 3. Dismantle each separately.
2   Unscrew the fuel filter cover screw and remove the screw, seal, filter cover and filter screen. Wash the screen in petrol.
3   Unscrew the float bowl nut and remove it, complete with sealing ring. Remove the float bowl and O-ring seal.
4   Unscrew the main jet, main jet holder and needle jet. Blow through the jets to ensure that they are clear.
5   Unscrew the starting jet and check the O-ring seal. Unscrew the pilot jet, and the accelerator pump valve. Blow through to clear.
6   Pull out the float pivot pin and remove the twin float. Drop out the float needle and unscrew the needle valve. Check the O-ring seal underneath the needle valve and inspect the float needle tip for wear. Check the plastic float for leaks by shaking to see if any petrol is inside.
7   Clean the float bowl and nut. Inspect the float chamber O-ring seal and fit it firmly in its groove before replacing the float bowl. Ensure that all O-ring seals are replaced without damage.
8   Unscrew the accelerator pump jet cover with its O-ring seal - this is opposite the fuel filter - and remove the accelerator pump jet with its O-ring seal. Blow through the jet. It should not be necessary to dismantle the accelerator pump. Do not move the pump adjuster screw - this is set by the factory.
9   Check the condition of the O-ring seal in the mixing chamber top. Inspect the throttle needle and needle jet for wear. The jet may be worn oval after considerable mileage when it should be renewed. The needle clip must be in groove 2.
10  Jet sizes and needle position are selected by the manufacturers after testing with the recommended fuel. Changes are only necessary in exceptional circumstances.
11  Do not move the throttle stop or mixture regulating screws or the carburettor will have to be retuned.
12  To remove the throttle slides and needles from the carburettor tops, pull back the return spring with one hand and invert the slide so that the throttle needle and clip can be pushed out. With the spring compressed, detach the slide from the cable. Control of the throttle slides by the main control cables is transmitted via a crank arrangement and a short length of cable within the carburettor top. Wear in this area is unusual. A worn cross shaft seal

may promote a weak mixture. The seal can be renewed after sliding the shaft from position.

13 Throttle slides or mixing chamber tops should not be interchanged as the accelerator pumps will then need readjusting.

14 Insert the throttle slide into the mixing chamber carefully, making sure that the needle goes into the needle jet. The accelerator pump lever is on the intake side.

## 6 Float chamber fuel level - checking

1 Remove both carburettors, see Section 3.

2 Detach the float bowl from each carburettor by unscrewing the central nut. Using a vernier gauge, check that the distance between the upper edge of the float and the float chamber mating surface is as follows, with the needle valve closed.

*18.5 mm (0.7283 in) with 10 gm floats*
*24.5 mm (0.9645 in) with 14 gm floats*

The floats are weight marked for easy identification. The float tongue may be bent carefully to make the necessary adjustment.

3 An incorrectly set float height can cause consistent overrichness of the mixture or weakness of the mixture depending upon whether the float is set too high or too low.

4.2a Remove plug to detach float chamber

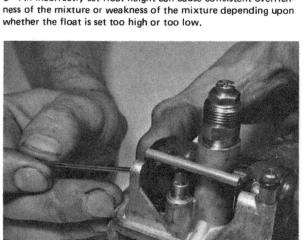
4.2b Displace pivot pin and lift float away

4.2c Float needle is located in fork

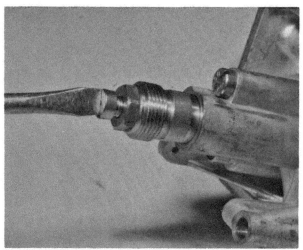
4.3a Unscrew the main jet and then ...

4.3b ... remove the accelerator pump body as a unit

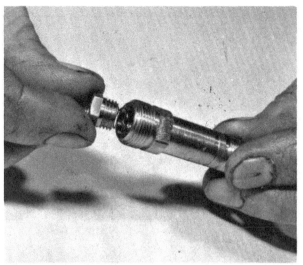

4.3c Unscrew the end plug to ...

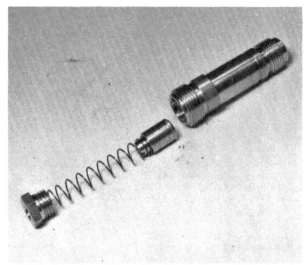

4.3d ... gain access to the pump components

4.3e Push down the main nozzle and ...

4,3f ... remove it from the bore

4.4a Unscrew the pilot jet and ...

4.4b ... remove the starter jet and 'O' ring

4.5 Unscrew the pilot mixture screw to aid cleaning

Fig. 3.3. Measuring float height

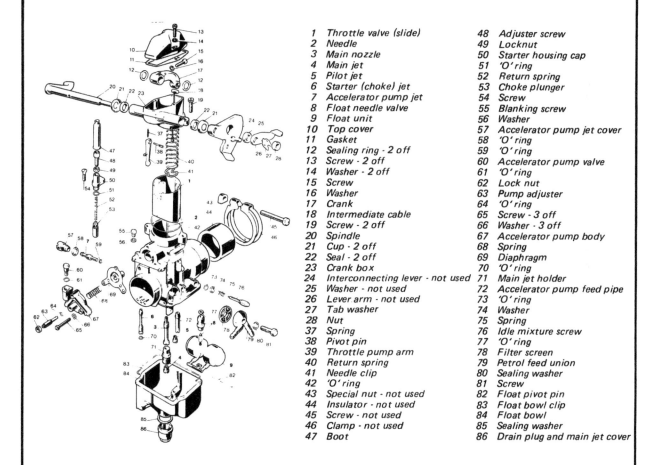

| | | | |
|---|---|---|---|
| 1 | Throttle valve (slide) | 48 | Adjuster screw |
| 2 | Needle | 49 | Locknut |
| 3 | Main nozzle | 50 | Starter housing cap |
| 4 | Main jet | 51 | 'O' ring |
| 5 | Pilot jet | 52 | Return spring |
| 6 | Starter (choke) jet | 53 | Choke plunger |
| 7 | Accelerator pump jet | 54 | Screw |
| 8 | Float needle valve | 55 | Blanking screw |
| 9 | Float unit | 56 | Washer |
| 10 | Top cover | 57 | Accelerator pump jet cover |
| 11 | Gasket | 58 | 'O' ring |
| 12 | Sealing ring - 2 off | 59 | 'O' ring |
| 13 | Screw - 2 off | 60 | Accelerator pump valve |
| 14 | Washer - 2 off | 61 | 'O' ring |
| 15 | Screw | 62 | Lock nut |
| 16 | Washer | 63 | Pump adjuster |
| 17 | Crank | 64 | 'O' ring |
| 18 | Intermediate cable | 65 | Screw - 3 off |
| 19 | Screw - 2 off | 66 | Washer - 3 off |
| 20 | Spindle | 67 | Accelerator pump body |
| 21 | Cup - 2 off | 68 | Spring |
| 22 | Seal - 2 off | 69 | Diaphragm |
| 23 | Crank box | 70 | 'O' ring |
| 24 | Interconnecting lever - not used | 71 | Main jet holder |
| 25 | Washer - not used | 72 | Accelerator pump feed pipe |
| 26 | Lever arm - not used | 73 | 'O' ring |
| 27 | Tab washer | 74 | Washer |
| 28 | Nut | 75 | Spring |
| 37 | Spring | 76 | Idle mixture screw |
| 38 | Pivot pin | 77 | 'O' ring |
| 39 | Throttle pump arm | 78 | Filter screen |
| 40 | Return spring | 79 | Petrol feed union |
| 41 | Needle clip | 80 | Sealing washer |
| 42 | 'O' ring | 81 | Screw |
| 43 | Special nut - not used | 82 | Float pivot pin |
| 44 | Insulator - not used | 83 | Float bowl clip |
| 45 | Screw - not used | 84 | Float bowl |
| 46 | Clamp - not used | 85 | Sealing washer |
| 47 | Boot | 86 | Drain plug and main jet cover |

Fig. 3.2. Carburettor - Le Mans model only

**Note:** The carburettor illustrated is similar to but not identical to the instrument fitted to the Le Mans model

## 7 Carburettors: idle speed adjustment and synchronisation

1 Before any running adjustments are made to the carburettors, the engine must be allowed to reach normal working temperature. In addition, it is important that the valve clearances and ignition timing are checked first.
2 With the engine stopped, screw the idle adjuster screws (pilot mixture screws) inwards and then outwards the number of turns prescribed in the Specifications.
3 Start the engine and screw the throttle adjuster screws in an equal amount until the engine is running at 1000-1200 rpm. Remove one plug cap and by means of the idle adjuster screw on the opposite carburettor find the point at which the engine runs fastest. If the screw was set correctly, the optimum position will not be far away. Replace the plug cap and repeat the operation on the other cylinder. Because of the engine configuration running on one cylinder can be accomplished only at relatively high engine speeds. This is why the initial engine speed must be set at 1000-1200 rpm.
4 When the mixture adjustment is correct, unscrew the throttle adjuster screw on one cylinder so that when the plug cap is removed the engine fires four or five times before stopping. Repeat this on the second cylinder. The correct idling speed should now be achieved.
5 In order to ensure correct synchronisation of the carburettors, it will be necessary to check that both carburettor slides start lifting at exactly the same time. Checking may be done visually, through the carburettor mouth. Adjustments should be made by means of the cable adjusters on the carburettor tops or at the twist grip control on the handlebars. After making the adjustment, screw the adjusters in or out an equal amount until there is approximately 3 mm (1/8 in) free play at the cables before throttle slide lift commences.
6 In addition to manual adjustments of the carburettors, the carburettors are fitted with take-off points to which vacuum gauges may be fitted. Follow the makers instructions for adjusting the carburettors, commencing at an engine speed of 800-900 rpm.

## 8 Air filter: removal and replacement. Except Le Mans models

1 To gain access to the air filter box and make removal of the air filter element easier, the petrol tank should be removed, followed by the carburettors as described in Section 3 of this Chapter. In addition, the battery must be removed as described in Chapter 7, Section 2.
2 Remove the rubber ducting from the rear of the breather box and detach the four hoses leading to the breather box. The duct is retained by a strap arrangement consisting of two thin steel stops and two springs.
3 Unscrew the single nut from the rod projecting through the front of the air filter box and withdraw the breather box followed by the air filter element.
4 The air filter element is of the dry paper type and should be renewed, irrespective of condition, at the recommended interval. Knock the element to remove the loose dust. Blow out from the inside with an air hose to remove the more ingrained matter. If the filter element is perforated or has become soiled with oil, it must be renewed. Poor performance and an over-rich mixture will result from a blocked filter.
5 When refitting the filter assembly note that the free plate at the front of the breather box must be located correctly with the projection on the box.

## 9 Oil filters: removal and cleaning

1 All models are fitted with a gauze mesh oil strainer within the sump, which should be cleaned every 9,000 miles (15,000 km). This is approximately every fifth oil change. With the exception of the 750S and most 850T models, a filter cartridge is also

fitted. The cartridge cannot be cleaned and must therefore be renewed at the same time as the screen is cleaned.
2 To gain access to the filter(s), drain the engine oil when the engine is warm and then remove the sump screws. The oil filters and pressure release valve are all attached to the sump plate itself.
3 Unscrew the filter cartridge and discard it. Bend down the lock plate on the gauze screen centre bolt. Remove the bolt and lift the screen from position. The screen may be cleaned in petrol and then allowed to dry thoroughly or blown dry with an air hose. When refitting the gauze, ensure that the centre bolt is secured by the tab washer.

## 10 Oil pump: removal, examination and renovation

1 The oil pump is of the spur gear type, retained in a housing fitted to the front of the engine and driven by the cam drive chain. Removal and inspection of the oil pump is unlikely to be necessary unless the engine lubrication system fails or during a major overhaul. The oil pump may be removed as described in Chapter 1, Section 10, after the frame has been separated from the engine and the alternator and timing cover removed. Refer to the relevant Sections in Chapter 1.
2 After withdrawing the Woodruff key from the oil pump driveshaft, push the integral shaft and drive pinion from position, and then displace the driven pinion. Clean the components thoroughly in petrol and allow them to dry. Make a visual check of the pinion teeth for chipping or wear and the shafts for scoring. Insert the driveshaft into the pump body and check for play in the two needle roller bearings. If play is evident, or if the needle rollers are scored, the bearings should be pushed from position and renewed.
3 Check the dimension of the gear pinions and the housings in the pump body against those given in the Specifications. Worn components cannot be repaired and therefore must be substituted by new components.

## 11 Oil pressure relief valve

1 The oil pressure relief valve consists of a spring loaded plunger contained within a housing screwed into the sump, adjacent to the filter. The valve regulates the pressure of oil in the lubrication system and allows the oil to bypass the filter and continue circulating if the filter becomes clogged.
2 To dismantle the valve, leaving the housing in the casing, remove the centre bolt and lift off the seal, cup and spring. Invert the casing and allow the plunger to fall out. The valve is set on initial assembly to a pressure of 54-60 psi (3.8-4.2 kg cm$^2$) by means of shims placed between the spring and cap.
3 Clean all components and reassemble them in the same order.

7.2 A = Throttle adjuster screw; B = Pilot mixture screw

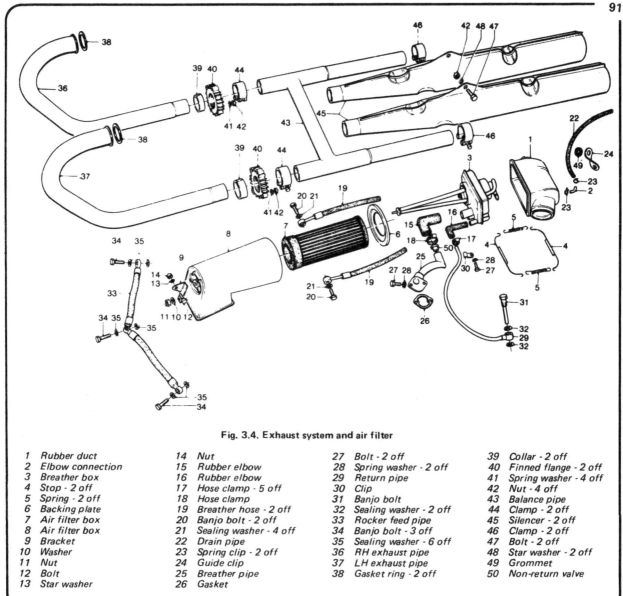

**Fig. 3.4. Exhaust system and air filter**

| | | | | | | |
|---|---|---|---|---|---|---|
| 1 | Rubber duct | 14 | Nut | 27 | Bolt - 2 off | 39 Collar - 2 off |
| 2 | Elbow connection | 15 | Rubber elbow | 28 | Spring washer - 2 off | 40 Finned flange - 2 off |
| 3 | Breather box | 16 | Rubber elbow | 29 | Return pipe | 41 Spring washer - 4 off |
| 4 | Stop - 2 off | 17 | Hose clamp - 5 off | 30 | Clip | 42 Nut - 4 off |
| 5 | Spring - 2 off | 18 | Hose clamp | 31 | Banjo bolt | 43 Balance pipe |
| 6 | Backing plate | 19 | Breather hose - 2 off | 32 | Sealing washer - 2 off | 44 Clamp - 2 off |
| 7 | Air filter box | 20 | Banjo bolt - 2 off | 33 | Rocker feed pipe | 45 Silencer - 2 off |
| 8 | Air filter box | 21 | Sealing washer - 4 off | 34 | Banjo bolt - 3 off | 46 Clamp - 2 off |
| 9 | Bracket | 22 | Drain pipe | 35 | Sealing washer - 6 off | 47 Bolt - 2 off |
| 10 | Washer | 23 | Spring clip - 2 off | 36 | RH exhaust pipe | 48 Star washer - 2 off |
| 11 | Nut | 24 | Guide clip | 37 | LH exhaust pipe | 49 Grommet |
| 12 | Bolt | 25 | Breather pipe | 38 | Gasket ring - 2 off | 50 Non-return valve |
| 13 | Star washer | 26 | Gasket | | | |

10.1 Oil pump secured by four screws, located on two dowels

10.2a Check the pump gears for scoring and ...

10.2b ... the driven shaft for wear

10.2c The double row roller bearing can be pushed out

11.2a Remove the cap bolt complete with the cap

11.2b Lift out the spring and valve plunger

### 12 Oil pressure warning switch

1   An oil pressure switch screwed into the crankcase forward of
the cylinders, and connected to a warning light, is provided to
indicate when the oil pressure has dropped below a safe level.
2   If the warning light illuminates when the engine is running
at anything over tick-over speed, a fault in the lubrication system
is indicated. Stop the engine immediately and do not restart
until the fault has been rectified.

### 13 Petrol level switch: location. V-1000 Convert model only

1   A float-operated switch is fitted in the underside of the
petrol tank, which gives a visual indication of low fuel level via
a warning lamp. Unless the float sticks in the cylinder in which
it moves, failure of the switch to operate is unlikely.
2   After draining the petrol from the tank the complete unit
may be unscrewed. The two wires leading to the switch are a
push fit on their terminals.

12.2 Oil pressure switch screws into crankcase housing

13.2a Two-piece float operated fuel level switch

13.2b Check condition of the sealing washer

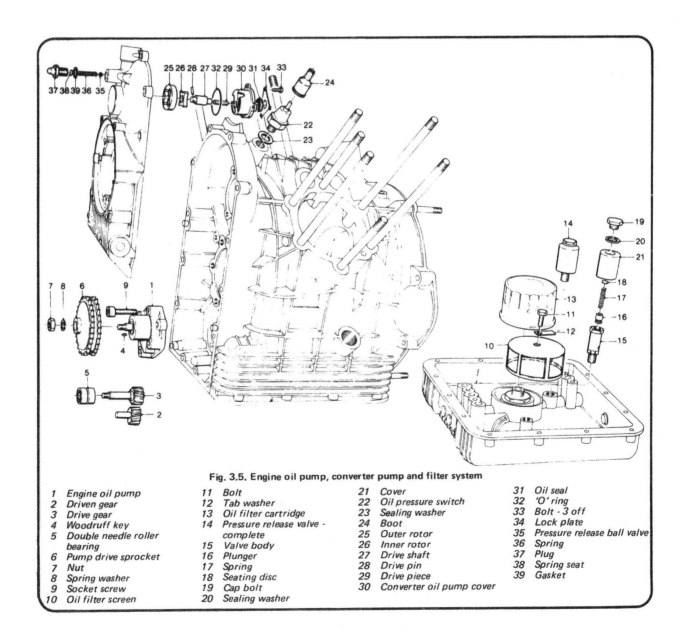

Fig. 3.5. Engine oil pump, converter pump and filter system

| | | | | | | | |
|---|---|---|---|---|---|---|---|
| 1 | Engine oil pump | 11 | Bolt | 21 | Cover | 31 | Oil seal |
| 2 | Driven gear | 12 | Tab washer | 22 | Oil pressure switch | 32 | 'O' ring |
| 3 | Drive gear | 13 | Oil filter cartridge | 23 | Sealing washer | 33 | Bolt - 3 off |
| 4 | Woodruff key | 14 | Pressure release valve - | 24 | Boot | 34 | Lock plate |
| 5 | Double needle roller | | complete | 25 | Outer rotor | 35 | Pressure release ball valve |
| | bearing | 15 | Valve body | 26 | Inner rotor | 36 | Spring |
| 6 | Pump drive sprocket | 16 | Plunger | 27 | Drive shaft | 37 | Plug |
| 7 | Nut | 17 | Spring | 28 | Drive pin | 38 | Spring seat |
| 8 | Spring washer | 18 | Seating disc | 29 | Drive piece | 39 | Gasket |
| 9 | Socket screw | 19 | Cap bolt | 30 | Converter oil pump cover | | |
| 10 | Oil filter screen | 20 | Sealing washer | | | | |

## 14 Fault diagnosis: carburation

| Symptom | Cause | Remedy |
|---|---|---|
| Engine 'fades' and eventually stops | Blocked air hole in filler cap | Clean. |
| | Empty fuel tank | Refill. |
| | Blocked fuel filter | Clean. |
| Engine difficult to start | Carburettor(s) flooding | Dismantle and clean carburettor(s). Check for punctured float. |
| | Twist grip open too far | Close twist grip. |
| | Air filter blocked | Remove and clean or renew. |
| Engine idles poorly | Blocked idling jet | Dismantle carburettor, and clean. |
| | Carburettor too rich or too weak | Adjust. |
| | Leak between carburettor and cylinder head | Check hose clips for tightness. |
| Engine runs badly - black smoke from exhausts | Carburettor(s) flooding | Dismantle and clean carburettor. Check for sticking float. |
| | Blocked air filter | Remove and clean or replace. |
| Engine difficult to start. Fires only occasionally and spits back through carburettors | Weak mixture | Check for fuel in float chambers and whether choke is closed. |
| Engine runs very hot, keeps on running after ignition is turned off | Weak mixture | Readjust carburettors. |
| Engine pinks | Fuel octane rating too low | Use correct grade fuel. |

## 15 Fault diagnosis: lubrication system

| Symptom | Cause | Remedy |
|---|---|---|
| Oil pressure warning lamp remains on, flashes or comes on when driving | Low oil pressure | Check. |
| | Filter blocked | Renew. |
| | Low oil level | Top-up. |
| | Worn main or big-end bearings | Dismantle and check. |
| | Faulty oil pump | Check. |
| | Faulty oil pressure relief valve | Check. |
| | Faulty oil pressure switch | Check continuity, renew if necessary. |
| | Faulty wiring | Check electrics. |
| White smoke in exhaust | Worn cylinder bore | Rebore. |
| | Broken piston ring | Renew. |
| | Worn valve guides | Renew. |

# Chapter 4 Ignition system

## Contents

## Specifications

### Contact breaker

| | |
|---|---|
| Make ... ... ... ... ... ... ... ... ... | Marelli |
| Type ... ... ... ... ... ... ... ... ... | S311A |
| Gap: | |
| 850T ... ... ... ... ... ... ... ... | 0.42-0.48 mm (0.016-0.018 in) |
| All others ... ... ... ... ... ... ... | 0.37-0.43 mm (0.014-0.017 in) |

### Ignition timing: BTDC

| | |
|---|---|
| 750S and S3 ... ... ... ... ... ... ... | 13° ± 1° retarded, 39° ± 3° advanced |
| 850T and Le Mans ... ... ... ... ... ... | 8° retarded, 34° advanced |
| 850T-3 ... ... ... ... ... ... ... | 2° retarded, 33° advanced |
| V-1000 I-Convert ... ... ... ... ... ... | 0°-2° retarded, 31°-33° advanced |

### Capacitor

| | |
|---|---|
| Type ... ... ... ... ... ... ... ... ... | CE 36 N |
| Capacity ... ... ... ... ... ... ... ... ... | 0.25 mfd |

### Ignition coil

| | |
|---|---|
| Make ... ... ... ... ... ... ... ... ... | Marelli |
| Type ... ... ... ... ... ... ... ... ... | BM 200C |
| Resistance at 20°C: | |
| Primary winding . ... ... ... ... ... ... ... | 3.35 ohms ± 6% |
| Secondary winding ... ... ... ... ... ... ... | 6.200 ohms ± 10% |

### Spark plugs

Gap ... ... ... ... ... ... ... ... ... Standard 0.6 mm (0.023 in), high speed 0.5 mm (0.019 in)

Type:

| | Standard - normal touring *Marelli | Standard - high speed *Marelli | Alternative Motorcraft |
|---|---|---|---|
| 750S and S3 ... ... ... ... ... ... ... ... | CW240L | CW275L | AG1 |
| 850T ... ... ... ... ... ... ... ... ... | CW240L | CW275L | AG23 |
| 850T-3 ... ... ... ... ... ... ... ... | CW7L | — | AG2 |
| V-1000 I-Convert ... ... ... ... ... ... | CW7L or CWLP | — | — |
| 850 Le Mans ... ... ... ... ... ... ... | Champion N9Y | — | — |

*original equipment

## 1  General description

The ignition system fitted to the Moto Guzzi V-twin models consists of two duplicated circuits, each of which serves one or other of the two cylinders.

The two contact breaker units are contained within a housing between the cylinders, and are operated by a single cam driven from a gear pinion on the valve camshaft. Low tension current passing through the ignition coils is interrupted when the contact breakers separate and by mutual induction create a high tension current in the ignition coil secondary windings, which is delivered to the spark plugs.

The ignition coils are mounted below the petrol tank under the frame top tube.

An automatic ignition timing unit (ATU) is fitted below the contact breaker operating cam, which alters the point of contact breaker separation, and hence the moment the spark occurs, to suit the engine speed.

A condenser is fitted to each ignition circuit, connected in parallel with the contact breaker to minimise arcing across the points faces and so maintain the high tension current.

The starter motor is a series-wound pre-engaged unit, which operates on direct current from the battery. A solenoid mounted on the top of the starter motor engages the starter pinion with the flywheel mounted ring gear. On all but Le Mans models, a switch incorporated in the clutch cable prevents operation of the starter motor unless the clutch is disengaged. V-1000 models are fitted with a switch which prevents operation of the ignition system with the side-stand in the extended position.

## 2  Contact breakers: adjustment

1  In order to inspect and adjust the contact breakers, the housing cap, retained by two screws must be removed. To aid access for easy removal and subsequent attention the rear of the petrol tank should be raised a few inches after detaching the retaining strap and the petrol pipes. Support the tank on a bunch of rags or a wooden block.

2  Rotate the engine until one set of points is open and examine the contact faces. Slight irregularities in the faces may be removed using a fine Swiss file or a strip of emery paper backed by a piece of tin. If they are dirty, pitted or burnt, it will be necessary to remove them for further attention, as described in Section 3 of this Chapter. Repeat the process for the second set of points.

3  The correct contact breaker gap, when the points are in the full open position, is within the range 0.42-0.48 mm (0.016-0.018 in) for 850T models and 0.37-0.43 mm (0.014-0.017 in) for all other models. Adjustment is effected by slackening the two screws passing through the contact breaker fixed point plate and using a screwdriver inserted in the notch provided, moving the fixed contact nearer to or further away from the moving contact. Ensure that the points are in the fully open position when this adjustment is made or a false setting will result. Tighten the two screws and recheck the gap; the feeler gauge should be a light sliding fit between the faces.

4  Repeat the process on the other set of contact points. Before refitting the housing cap, clean the points faces using a clean rag dipped in methylated spirits. This will ensure that the points are perfectly clean and prevent the faces picking-up prematurely. Apply a few drops of thin oil to the cam lubricator wick. Do not over lubricate or the excess oil may find its way to the points, causing ignition failure.

## 3  Contact breakers: removal, renovation and replacement

1  If the contact breaker points are burned, pitted or badly worn, they should be removed for dressing. If it is necessary to remove a substantial amount of material before the faces can be restored, new contacts should be fitted.

2  To remove a contact breaker set, loosen the screw retaining the leads from the condenser and low tension circuit and pull the forked terminals from place. Remove the two screws passing through the fixed point baseplate and lift the complete contact breaker unit from position.

3  To separate the moving contact from the fixed contact plate remove the terminal screw completely and slide off the plastic insulator and cage to release the return spring end. Prise off the E clip from the top of the pivot post and remove the moving point from the pivot.

4  The points should be dressed with an oilstone or fine emery cloth. Keep them absolutely square during the dressing operation, otherwise they will make angular contact when they are replaced and will burn away rapidly as a result.

5  Replace the contacts by reversing the dismantling procedure, taking care to position the insulating washers in the correct sequence. Lightly grease the pivot post before replacing the moving contact and check that there is no oil or grease on the surface of the points. Place a few drops of oil on the lubricating wick that bears on the contact breaker cam, so that the surface is kept lubricated.

6  Readjust the contact breaker gap to the recommended setting, after verifying that the points are in their fully open position.

7  Repeat the whole procedure for the other set of contact breaker points.

## 4  Condensers: location, removal and replacement

1  Condensers are included in the contact breaker circuit to prevent arcing across the contact breaker points as they separate. A condenser is connected in parallel with each set of contact points, and if a fault develops in either, or both condensers, ignition failure is liable to occur.

2  If the engine is difficult to start, or if misfiring occurs, it is possible that a condenser is at fault. To check whether a condenser has failed, observe the points whilst the engine is running. If excessive sparking occurs across one set of points and they have a blackened or burnt appearance, it may be assumed the condenser in that circuit is no longer serviceable.

3  The condensers are mounted on the outside of the contact breaker housing. Each is retained by a single screw through the strap soldered to the body of the condenser and by the lead wire attached by the screw passing through the end of the moving contact return spring.

Loosen the screw slightly so that the forked terminal on the condenser wire can be detached.

4  Because it is impracticable to repair a defective condenser, a new one must be fitted. Note that it is extremely unlikely that both condensers will fail in unison; if total ignition failure occurs the source of the trouble should be sought elsewhere.

2.1 Raise the petrol tank for contact-breaker access

2.3 Point gap adjustment screws

3.2a Remove both screws and detach wires ...

3.2b ... to free complete contact-breaker unit

3.3a Prise off the 'E' clip to remove ...

3.3b ... the moving contact

4.3a Condensers are retained by a single screw

4.3b Remove clip to free condenser wires

## 5  Condensers: testing

1  Without the appropriate test equipment, there is no alternative means of verifying whether a condenser is still serviceable.
2  Bearing in mind the low cost of a condenser, it is far more satisfactory to check whether it is malfunctioning by direct replacement.

## 6  Ignition coils: checking

1  Each cylinder has its own ignition circuit and if one cylinder misfires, one half of the complete ignition system can be eliminated immediately. The components most likely to fail in the circuit that is defective are the condenser and the ignition coil since contact breaker faults should be obvious on close examination. Replacement of the existing condenser will show whether the condenser is at fault, leaving by the process of elimination the ignition coil.
2  A suspect ignition coil can be tested completely only using specialist equipment. It is possible, however, to gain some indication of the coil's condition as follows: Remove the contact breaker housing cap and the spark plug cap from the circuit in question. Switch the ignition on and turn the engine over until the points are fully closed. Hold the uninsulated HT cable end about 3.0 mm (1/8 in) away from a suitable part of the cylinder head. Using a screwdriver with an insulated handle 'flash' the contact breaker points open. If the spark produced at the HT lead end is able to jump a gap of 3.0-6.0 mm (1/8 - 1/4 in) it is probable that the coil is in good condition.
3  The ignition coils are sealed units and it is not possible to effect a satisfactory repair in the event of failure. A new coil must be fitted.
4  The two coils are fitted underneath the petrol tank, below the frame top tube, and are supported by  straps and bolts. After removal of the tank disconnect the HT leads by unscrewing the connector ferrules and pull the low tension wires from the terminals. Note the position of the wires for easy reassembly. The coils will come away after removing the strap bolts.

## 7  Ignition timing: checking and resetting

1  The ignition timing should be checked at the period stated in the routine maintenance schedule or when the contact breakers have been adjusted or renewed. Do not retime the ignition unless the contact breaker points gaps have been checked first and if necessary readjusted.

2  Raise the rear of the petrol tank and remove the contact breaker cap as described in Section 2 of this Chapter. Detach the spark plug leads and remove the spark plugs. Remove the alternator cover, to gain access to the alternator centre bolt which can be used after fitting a socket key to rotate the engine. Prise the rubber inspection plug from the right-hand side of the gearbox housing so that the timing index marks may be viewed through the aperture.
3  The ignition timing for the right-hand cylinder should be checked first. To establish the moment at which the contact breaker points open, connect one lead from a lamp to the red lead of the contact breaker points and the other lead to a suitable earth point on the engine. A multi-meter or buzzer unit may be used in place of the bulb arrangement. Rotate the engine clockwise until the right-hand cylinder is on the compression stroke. This can be felt by placing a thumb over the spark plug hole. Rotate the engine further until the retarded timing mark for the right-hand cylinder (see Fig. 4.1 for identification) aligns exactly with the gearbox housing aperture index mark. The indicator lamp should illuminate at the moment the two lines coincide exactly.
4  If the indicator light illuminates *before* the mark aligns, the ignition is too far advanced, and if the light comes on *after* the two marks align, the ignition is too far retarded. Rotate the engine back again and then clockwise once more until the marks align. Always rotate the engine in a clockwise direction when bringing the marks into line, to take up any backlash in the drive gear.
     To adjust the timing on the right-hand cylinder, loosen the two bolts which retain the contact breaker housing clamp and rotate the housing until the contact breaker points open at exactly the right moment. After adjustment, tighten the screws and recheck by repeating the timing check procedure.
     The two bolts holding the housing in place are partially obscured by the housing and the right-hand cylinder. To aid access, a special tool no 14-92-70-00 can be obtained. If this is not available, an existing spanner could be modified especially for this purpose.
5  The left-hand cylinder ignition timing can be checked using the same procedure after connecting an indicator lamp or buzzer to the left-hand contact breaker (green wire). Rotate the engine until the retarded timing mark for the left-hand cylinder (see Fig. 4.1 for identification) aligns exactly with the gearbox housing aperture index mark. If the timing is incorrect, loosen the two screws which secure the contact breaker assembly baseplate and rotate it until the timing is correct. Tighten the screws and recheck.
6  Having checked and reset the ignition timing with the engine at rest (static timing) the ignition timing should then be checked by using a stroboscope, with the engine running. This will enable the correct functioning of the automatic advance unit to be checked. Prior to the check it is advisable to apply a trace of white paint to the retarded and full advance marks (see Fig. 4.1) as they will then be more easily seen when using the stroboscope.
     Connect the strobe lamp up to the right-hand cylinder by following the manufacturer's instructions. Start the engine and allow it to run at tick-over. Aim the stroboscope at the flywheel through the viewing aperture. If the timing is correct, the marks should align as described in the static test. Raise the engine speed progressively to 6000-6200 rpm when the full advance mark should appear in the aperture. If the ignition does not advance correctly, the automatic advance unit should be checked.

## 8  Automatic advance unit: dismantling and examination

1  If loss of power or engine roughness has been experienced, the automatic advance unit (ATU) should be examined. The ATU may be inspected with the contact breaker housing in place on the engine, after removal of the contact breaker assemblies. Having detached the housing cap, mark the relative

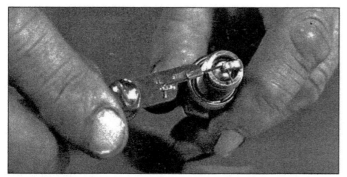

**Electrode gap check** - use a wire type gauge for best results

**Electrode gap adjustment** - bend the side electrode using the correct tool

**Normal condition** - A brown, tan or grey firing end indicates that the engine is in good condition and that the plug type is correct

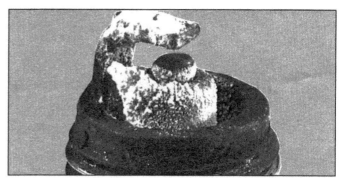

**Ash deposits** - Light brown deposits encrusted on the electrodes and insulator, leading to misfire and hesitation. Caused by excessive amounts of oil in the combustion chamber or poor quality fuel/oil

**Carbon fouling** - Dry, black sooty deposits leading to misfire and weak spark. Caused by an over-rich fuel/air mixture, faulty choke operation or blocked air filter

**Oil fouling** - Wet oily deposits leading to misfire and weak spark. Caused by oil leakage past piston rings or valve guides (4-stroke engine), or excess lubricant (2-stroke engine)

**Overheating** - A blistered white insulator and glazed electrodes. Caused by ignition system fault, incorrect fuel, or cooling system fault

**Worn plug** - Worn electrodes will cause poor starting in damp or cold weather and will also waste fuel

positions of the contact breaker assembly baseplate and the housing body with a centre punch, so that the plate may be refitted in exactly the same position for ease of retiming.

2   Detach the two condensers, each of which is secured by a single screw, and after removing the two baseplate screws lift off the complete assembly. Grasp the contact breaker operating cam and rotate it to left and right, checking that it moves smoothly and the bob weights open and shut against the pressure of the two return springs. The heavier of the two springs does not come into operation until the bob weights are approximately half open.

3   To dismantle the unit, prise the outer ends of the springs off the anchor pins on the bob weights. Lift the lubricator wick from position in the end of the cam and remove the centre screw. Before lifting the cam from place, note the position of the cam relative to the housing body. The cam must be refitted in the same position to maintain the ignition timing.

Tightness in the ATU mechanism can usually be overcome by lubrication. If the ends of the springs are damaged or the springs appear weakened, both springs should be renewed to restore the original advance-retard characteristics of the unit. If the cam appears worn or the bob weights and pivots are worn, the ATU complete with the contact breaker housing must be renewed as a unit. The component parts are not available as separate items.

4   Reassemble the unit, applying light engine oil to the working surfaces.

7.3a Timing mark inspection aperture; 'S' mark is TDC on LH cylinder

7.3b RH cylinder timing; clamp bolts

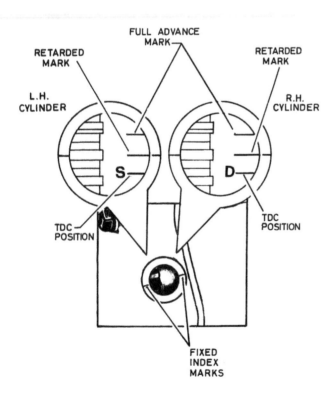

Fig. 4.1. Ignition timing marks

7.3c LH cylinder timing; base plate screws

8.3a Remove centre screw and ...

8.3b... lift cam unit from the ATU

8.3c Check the condition of the bob-weights and springs

## 9 Contact breaker housing: removal and replacement

1   Removal of the contact breaker housing can take place only after the engine/gearbox unit is free from the frame or after the frame has been lifted up sufficiently so that the housing will clear the right-hand frame top tube as it is withdrawn.

2   Before removing the contact breaker housing, place the engine at TDC and mark the relative positions of the following components: crankcase, housing, contact breaker, baseplate and cam. If, on reassembly, the engine is again placed at TDC, and the punch marks are realigned, the contact breaker housing may be refitted with greater ease, without having to resort to retiming the camshaft and contact breaker cam driveshaft relationships.

3   To remove the housing, loosen both clamping bolts and then remove the forward bolt. Swing the clamp around to clear the housing boss and pull the housing from position. With the exception of the contact breaker components and the ATU springs, the only part of the contact breaker housing that may be renewed as a separate item is the driven gear pinion on the lower end of the shaft. The pinion is secured by a roll pin passing through the pinion boss and the shafts. The pin maybe drifted from place and a new pinion fitted. If wear of the pinion has developed, it is probable that the drive gear on the camshaft is also badly worn.

If this is the case, the two components should be renewed as a pair.

4   Refit the contact breaker housing by reversing the dismantling procedure. If alignment marks were not made on dismantling, replace the housing as follows: Place the engine so that the right-hand cylinder is on the compression stroke and the static timing mark is aligned with the aperture mark on the gearbox bellhousing. Rotate the contact breaker camshaft in a clockwise direction (viewed from above) until the right-hand cylinder points are just opening. Hold the cam in this position and insert the housing into the crankcase so that the drive gears mesh. The housing must be replaced so that the right-hand cylinder contact breaker set (red wire) is on the left-hand side of the machine. Fit the clamp and partially tighten the two bolts. The ignition timing should be carried out as described in Section 7 of this Chapter.

## 10 Starter motor: removal

1   Note the positions of the leads running to the starter motor and then remove them. As when dismantling or detaching any electrical component the battery should be disconnected first.

2   The starter motor is retained on the engine casing by two bolts. Support the weight of the motor and remove the bolts. Lift the motor out towards the rear.

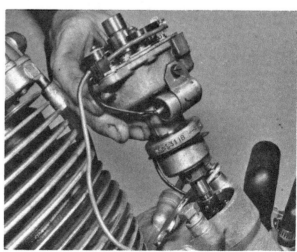

9.3a Mark the housing before removal to aid reassembly

9.3b The driven gear is retained by a pin

10.1 Pre-engagement type starter motor

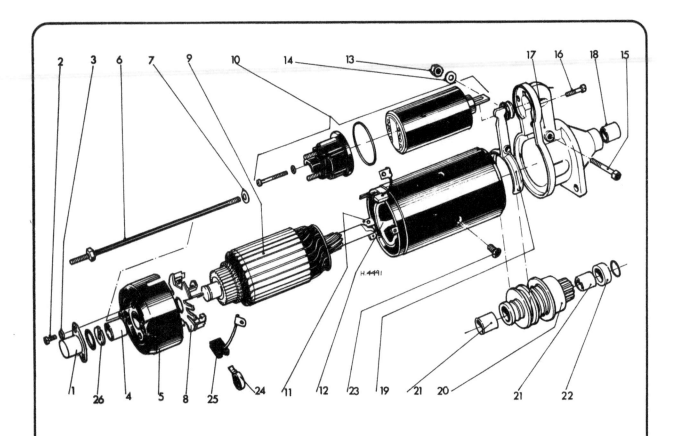

Fig. 4.2. Starter motor - except 750S model

| | | | | | | |
|---|---|---|---|---|---|---|
| 1 | End cap - 1 off | 7 | Spring washer M5 - 2 off | 13 | Hex. nut - 1 off | 20 | Starter pinion - 1 off |
| 2 | Screw - 2 off | 8 | Brush carrier - 1 off | 14 | Spring washer - 1 off | 21 | Bush - 1 off |
| 3 | Spring washer M4 - 2 off | 9 | Armature - 1 off | 15 | Pivot bolt - 1 off | 22 | Stop ring - 1 off |
| 4 | Bush - 1 off | 10 | Starter solenoid - 1 off | 16 | Countersunk screw - 1 off | 23 | Countersunk screw - 1 off |
| 5 | Commutator cover - 1 off | 11 | Insulating strip - 1 off | 17 | Drive housing - 1 off | 24 | Brush spring - 4 off |
| 6 | Starter stud - 2 off | 12 | Field coil - 1 off | 18 | Bronze bush - 1 off | 25 | Brush set - 1 off |
| | | | | 19 | Lever - 1 off | 26 | Circlip - 1 off |

## 11 Starter motor: overhaul

1   Two entirely different types of starter motor are fitted to the range of Moto Guzzi models covered in this manual. All 750S models are fitted with a Bosch inertia type starter motor, the pinion of which is engaged on the Bendix principle. In this system the pinion is thrown into engagement with the starter ring gear on the flywheel by the rotating armature. All later models have pre-engagement starter motors, where the pinion is moved into mesh by a pivoted arm operated by a solenoid mounted on the starter motor body. In this way the two gears may be fully meshed before the starter motor starts rotating.

### Pre-engagement type

2   To renew the brushes, remove the two screws and washers which secure the small dust cap. Remove the cap and prise off the circlip and washer. Remove the two long screws which retain the starter motor end cap and then lift the cap away. Examine the brushes and check that they move freely. Sticking brushes will prevent correct functioning of the motor as electrical continuity between the brushes and the segmented commutator will be impossible.

3   Pull up the springs of the two brushes attached to the brush holder, partly withdraw the brushes and wedge them in raised position with the springs. Pull up the springs of the remaining two brushes attached to the field coils and remove the brushes completely. Remove the brush holder backplate.

4   If the brushes have worn to half their original length, they may be unsoldered. When renewing them, do not allow solder to run up the copper brush tails, towards the brushes.

5   Clean the commutator with fine glass paper, not emery. Ensure that the commutator segments are undercut, that is the insulation between each segment should be 0.5 mm (0.020 in) below the commutator surface. Use a hacksaw blade ground to the correct thickness to undercut the insulation. If the commutator is badly scored, it must be skimmed to a fine surface finish. Do not reduce the diameter below 33 mm (1.299 in). Note the washer and insulating washer on the armature shaft.

6   When refitting the commutator cover, position the brush holder backplate so that the long screws pass through the two slots in the edge of the plate.

7   To remove the starter pinion, first detach the field coil lead from the solenoid. Unscrew the retaining screws and remove the solenoid, disengaged from the pinion engaging lever.

8   Remove the brush holder plate as previously described. Pull the armature with pinion and pinion housing out of the starter body.

9   Unscrew the engaging lever pivot screw and take out the armature with engaging lever.

10   Push the thrust ring clear of the wire circlip and remove the circlip. Take off the starter pinion assembly.

11   A new pinion housing bush may be pressed into place, after soaking it in engine oil for 30 minutes. The end of the bush should be flush with the housing.

12   Inspect the field coils for scorching or charring. Electrical checking of the armature and field coils and renewal of the field coils, has to be left to a specialist.

13   Before reassembly, coat the quick thread and the engaging ring with the recommended grease. Check the armature end-float, which is adjusted by means of shims.

### Inertia type

14   After removal of the end cover and dust cover the starter motor may be inspected as described for the pre-engagement type of unit. Brushes worn below 11.5 mm (0.4527 in) should be renewed. When attending to the commutator it should be noted that the minimum permissible diameter is 31.2 mm (1.228 in).

15   To remove the pinion push the end collar back towards the starter motor, against the pressure of the pinion return spring. Prise out the circlip and then pull off the collar, spring and pinion.

## 12 Spark plugs: checking and setting the gap

1   Use only spark plugs of the correct reach and heat range, as recommended in the Specifications. Check the electrode gap at the recommended mileage, and when replacing plugs.

2   Use a plug spanner that is a good fit, otherwise an insulator may be broken. The plug should be tightened only sufficiently to provide a good gastight seal. If a plug is overtightened, the cylinder head threads could be damaged. Although they can be reclaimed with a 'Helicoil' wire thread insert, the cylinder head will need to be removed.

3   Check each plug gap with a feeler gauge. To reset, bend only the outer electrode. Clean the electrodes with a wire brush. Examine the insulation around the central electrode for cracks. Smear the threads with graphite grease before replacing. If the electrodes are badly eroded, the spark plug must be replaced.

4   The condition of the spark plug electrodes gives a good guide to engine operation. See accompanying diagrams.

5   Examine the spark plug caps and leads for cracks, scorching, or damage and ensure that they make good contact with the plug.

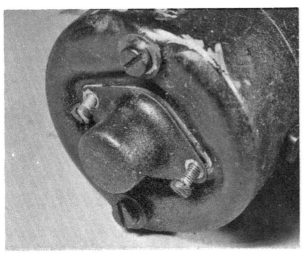

11.2a Remove the end cap and ...

11.2b ... prise off the circlip and washer

11.5 Check brush length and inspect commutator for scoring

11.7 Inspect starter pinion and flywheel ring gear teeth for damage

### 13 Fault diagnosis: ignition system

| Symptom | Cause | Remedy |
|---|---|---|
| Engine refuses to start | Break or short circuit in ignition system | Switch off and check wiring. |
| | Contact breaker not opening, or gap too large | Adjust contact breaker. |
| | Contact breaker points dirty | Clean. |
| | Plugs wet due to condensation or excess fuel | Remove and clean plugs. |
| | Plug gaps too wide | Adjust gap. |
| | Flat battery | Recharge battery. |
| Engine will not idle | Faulty capacitor | Renew capacitor. |
| | Incorrect timing | Adjust timing. |
| | Fouled or wrong grade plugs | Check or clean plugs. |
| | Centrifugal advance springs weak | Renew springs. |
| | Unit jammed on full advance | Free bob weights. |
| | Plug gaps too large or fouled | Adjust or clean plugs. |
| | Plug leads wet or defective | Dry or renew leads. |
| | Plug caps shorting (recognisable by sooty burn marks) | Dry or renew caps. |
| Engine lacks response, overheats, or runs on after ignition is switched off | Contact breaker points gap too small | Check points gap. |
| | Centrifugal advance jammed in retard position | Free bob weights. |
| | Timing retarded | Check timing. |
| | Wrong grade plugs | Renew plugs. |
| Engine 'pinks' under load | Pre-ignition; timing incorrect | Renew plugs and check timing. |

**14 Fault diagnosis: starter motor**

| Symptom | Cause | Remedy |
|---------|-------|--------|
| Starter fails to turn when button is pressed | Switch on lighting switch - if lights do not work, suspect flat battery or disconnected battery lead | Recharge battery.<br>Check leads and terminals. |
| | If lights fade, battery needs charging<br>If lights come on, but go out as soon as the starter is operated - suspect poor battery connections | Recharge battery.<br>Clean terminals and earth connection. |
| | Brushes too short or stuck<br>Inadequate brush pressure<br>Dirty commutator | Renew or release brushes.<br>Replace brush springs.<br>Clean commutator. |
| Starter will not turn engine | Commutator dirty<br>Faulty armature or field coil | Clean commutator.<br>Renew armature or coil. |
| Starter turns at high speed, but engine turns jerkily or not at all | Drive pinion worn<br>Flywheel spur gear worn<br>Drive pinion thread fouled or damaged | Renew drive pinion.<br>Renew flywheel.<br>Repair or clean thread. |

# Chapter 5 Frame and forks

## Contents

## Specifications

### Frame
| | |
|---|---|
| Type ... ... ... ... ... ... ... ... ... | Duplex cradle |

### Front forks
| | |
|---|---|
| Stanchion diameter ... ... ... ... ... ... ... | 34.715 - 34.690 mm (1.3667 -1.3657 in) |
| Lower leg internal diameter ... .. .. ... ... ... | 34.750 - 34.790 mm (1.3681 - 1.3696 in) |
| Stanchion/lower leg clearance ... ... ... ... ... ... | 0.045 - 0.100 mm (0.0017 - 0.004 in) |

Front fork oil capacity (per leg)

| Model | Quantity |
|---|---|
| 750S and 850T ... ... ... ... ... ... ... ... | 0.050 ltr (1.7/1.4 US/Imp oz) |
| 850T3 ... ... ... ... ... ... ... ... ... | 0.060 ltr (2.0/1.7 US/Imp oz) |
| 750S3 and V-1000 ... ... ... ... ... ... ... | 0.070 ltr (2.4/2.0 US/Imp oz) |
| Le Mans ... ... ... ... ... ... ... ... ... | 0.120 ltr (4.0/3.4 US/Imp oz) |
| Damper oil type ... ... ... ... ... ... ... ... | Dexron Ⓡ ATF |

Front fork spring free length ... ... ... ... ...
| | |
|---|---|
| V-1000 | |
| Part No. 14 52 66 00 ... ... ... ... ... ... | 421 ± 2.5 mm (16.575 ± 0.10 in) |
| Part No. 18 52 66 00 ... ... ... ... .. . ... | 415 ± 2.5 mm (16.339 ± 0.10 in) |
| All others ... ... ... ... ... ... ... ... | 418.500 - 423.500 mm (16.476 - 16.672 in) |

### Rear suspension
| | |
|---|---|
| Type ... ... ... ... ... ... ... ... ... | Swinging arm supported on hydraulic suspension units |
| Spring free length: | |
| V-1000 and 850T3 ... ... ... ... ... ... ... | 270 mm (10.6 in) |
| 750S and S3 ... ... ... ... ... ... ... ... | 277 ± 5 mm (10.9 ± 0.2 in) |
| 850T ... ... ... ... ... ... ... ... ... | 300 mm (11.8 in) |
| Le Mans .. ... ... ... ... ... ... ... ... | 279 mm (10.9 in) |

## 1 General description

The frame utilised on the Moto Guzzi models covered in this manual is of the duplex cradle type, fabricated from tubular members. The two lower horizontal members, to which the engine/gearbox unit and centre stand are bolted, are detachable from the main frame unit to allow easy engine removal.

The front forks are of the traditional telescopic type but utilise an unconventional method of damping. The hydraulic dampers are sealed units working independently from the oil contained in each fork leg, which acts only as a lubricating medium.

Rear suspension is provided by a swinging arm fork pivoting on tapered roller bearings, and supported on adjustable hydraulically-damped rear suspension units. The right-hand arm of the fork also serves as a shroud for the driveshaft and as a mounting for the bevel drive housing.

## 2  Front forks: removal from the frame

1    It is necessary to remove the complete fork assembly from the frame only when the steering head bearings require attention, or if the frame or forks have to be renewed due to damage.

2    Place the machine on the centre stand and position wooden blocks below the engine sump so that the front wheel is well clear of the ground. On touring models detach the handlebar screen by removing the mounting bolts. On Le Mans models, detach the cockpit fairing, taking care not to damage the seal around the headlamp rim.

3    Remove the petrol tank to prevent damage to the painted surface and to improve access to the steering head lug. Disconnect the battery to prevent short circuits occurring when removing the electrical leads to the instruments. Disconnect the speedometer cable, and tachometer cable (where fitted) by unscrewing the knurled securing rings.

4    Disconnect the wiring leads to the instruments and warning lamp console either at the instruments or at the snap connectors. Choose the most appropriate method for each wire. To gain access to the wiring within the headlamp shell remove the glass/reflector unit complete with the headlamp rim. The complete assembly is retained by a single screw, which passes through the base of the rim into the headlamp shell. Pull the socket from the rear of the headlamp shell and pull out the pilot bulb (where fitted) to detach the glass/reflector unit completely. The instruments and panel are retained on the fork upper yoke by two bolts or screws passing through rubber sleeves. Note the position of the sleeves and washers on removal. When detaching the wiring leads note carefully their original positions, to aid reassembly.

5    Detach the controls from the handlebars, making disconnections only where necessary. If cable or wire length permits, detach the controls and thread them through as necessary, so they may be placed out of the way to the rear of the machine. This will save additional work on reassembly. When removing the front brake master cylinder unit ensure that the reservoir cap is tight, and keep the reservoir upright, to prevent spillage of fluid. Hydraulic fluid is a very good paint stripper! Tie the master cylinder unit to a convenient frame part until the right-hand disc caliper is detached. The two components may be removed together without detaching the connecting hose and so necessitating rebleeding of the brakes on reassembly.

6    Except on Le Mans and 750S models, detach the handlebars

by removing the two half clamps held by two screws each.

7    On twin front disc models, remove one of the caliper units, by unscrewing the two main bolts. **Do not** disconnect the brake hose. Remove the nut and washer from the front wheel spindle. Loosen the two clamping screws which secure the spindle, and with the front wheel supported, withdraw the spindle. If necessary, pass a tommy bar through the hole in the spindle head to aid removal. Pull the wheel spacer from position and lower the wheel away from the front forks.

8    Remove the single bolt which secures the right-hand front brake switch junction to the lower yoke. Detach the brake hose clamp(s) from the mudguard. On machines with integrated braking, detach the hose guide from the left-hand side of the frame, below the head lug.

9    The front brake caliper is secured to the fork leg by two bolts. Remove the bolts and tie up the caliper so that it does not hang on the brake pipe or hose. Detach the mudguard after removing the bolts which pass into the fork legs. The caliper(s) may now be moved rearwards, the right-hand one being detached from the machine complete with the master cylinder, and the left-hand one being tied to a suitable part of the frame where it is away from further dismantling.

10   On Le Mans and 750S models, slacken the clamping screws which secure the separate handlebar 'clip-ons' to each fork leg. When the fork legs are moved downwards, the handlebar stubs may be removed from the stanchions.

11   Loosen the two clamp bolts which secure each fork leg to the yokes. The legs may now be removed from the yokes by gently easing them downwards. If necessary, aid removal by using a rawhide mallet.

12   On machines fitted with an hydraulic steering damper, free the damper at the fork end by removing the domed nut or bolt. Where an adjustable damper is utilised, detach the control quadrant from the lower end of the rod and withdraw the rod and steering damper knob.

13   Slacken the pinch bolt which passes through the rear of the fork upper yoke. Remove the sleeve nut from the top of the steering stem and lift the upper yoke from position. The headlamp shell, complete with fork shrouds, should be supported as the yoke is lifted away and then removed.

14   To remove the lower yoke and steering stem, unscrew the adjuster nut from the upper end of the steering stem and remove the dust cap. Lower the yoke downwards complete with the steering stem and steering head bearing lower inner race. The upper bearing will stay in the outer race.

2.7a Remove front wheel spindle nut and washer and ...

2.7b ... slacken off the fork clamp bolts

2.7c Pass a tommy bar through wheel spindle head to remove

2.7d Lower the wheel from position

2.9 Front mudguard stays held by two bolts each at rear

2.11a Slacken the upper and ...

2.11b ... lower stanchion securing pinch bolts

2.11c Ease the fork leg downwards and remove complete

### 3  Front forks: dismantling the fork legs

1   The front fork legs may be removed from the machine without dismantling the steering yokes or disturbing the instruments,
and headlamp, by following the procedure described in the
previous Section, paragraphs 7, 8, 9, 10 and 11. The fork legs
should be dismantled individually, to prevent the accidental
interchange of matched components. Before dismantling either
leg, remove the drain screw from the base of the lower leg and
allow the fluid to drain into a suitable container. This is not the
damping fluid but merely the lubricating oil.

2   Remove the socket screw from the centre of the stanchion
top bolt and then remove the bolt itself. If necessary, place the
stanchion in the jaws of a vice, protected by a piece of inner
tubing, to loosen the bolts. Invert the fork leg and remove the
socket screw which locates the end of the internal damper rod.

3   Prise off the dust cap from the fork lower leg and withdraw
the stanchion, complete with damper assembly and fork spring.
Pull the spring and damper from the stanchion. To remove the
spring from the damper rod, compress the spring slightly and
using a pair of circlip pliers displace the circlip from the fork
end. The spring guide and spring are then free for removal. The
spring seat into which the damper rod and spring fit can be
displaced from the inside of the fork lower leg by inverting the
leg and jarring it lightly on the workbench

3.1 Lubricate drain plug in the base of each leg

3.2a Damper unit seats in fork top bolt and is ...

3.2b ... retained by a socket screw

3.2c Unscrew socket bolt in lower leg

3.3a Separate stanchion from upper leg after ...

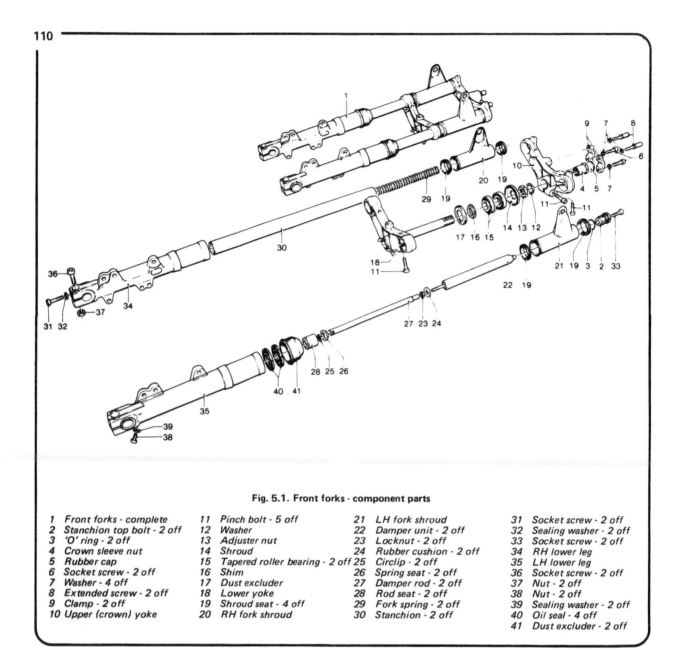

**Fig. 5.1. Front forks - component parts**

| | | | | | | |
|---|---|---|---|---|---|---|
| 1 | Front forks - complete | 11 | Pinch bolt - 5 off | 21 | LH fork shroud | 31 | Socket screw - 2 off |

1  Front forks - complete
2  Stanchion top bolt - 2 off
3  'O' ring - 2 off
4  Crown sleeve nut
5  Rubber cap
6  Socket screw - 2 off
7  Washer - 4 off
8  Extended screw - 2 off
9  Clamp - 2 off
10  Upper (crown) yoke

11  Pinch bolt - 5 off
12  Washer
13  Adjuster nut
14  Shroud
15  Tapered roller bearing - 2 off
16  Shim
17  Dust excluder
18  Lower yoke
19  Shroud seat - 4 off
20  RH fork shroud

21  LH fork shroud
22  Damper unit - 2 off
23  Locknut - 2 off
24  Rubber cushion - 2 off
25  Circlip - 2 off
26  Spring seat - 2 off
27  Damper rod - 2 off
28  Rod seat - 2 off
29  Fork spring - 2 off
30  Stanchion - 2 off

31  Socket screw - 2 off
32  Sealing washer - 2 off
33  Socket screw - 2 off
34  RH lower leg
35  LH lower leg
36  Socket screw - 2 off
37  Nut - 2 off
38  Nut - 2 off
39  Sealing washer - 2 off
40  Oil seal - 4 off
41  Dust excluder - 2 off

3.3b ... prising off the dust excluder

3.3c Remove the circlip and ...

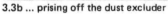

3.3d ... the spring lower seat to free ...

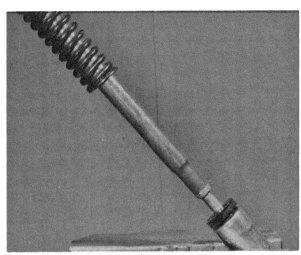

3.3e ... the fork spring from the damper rod

## 4  Front forks: examination and renovation

1   Check that the damper units are in good condition. Both should offer substantial resistance in both directions, the greater resistance being on extension. The damper units should give equal amounts of resistance to movement. If the damper units require renewal, they may be unscrewed from their rods after loosening the locknuts. Always replace the dampers as a pair.
2   Check the fit of the stanchions in the fork legs. If any lateral movement is evident, the worn components must be renewed, as no bushes are fitted. If the stanchions have become bent in an accident, they should be renewed. Although straightening is often possible the risk of subsequent fracture cannot be accepted.
3   The free length of each fork spring is given in the Specifications. If the springs appear weakened or are of different lengths, check them against a new spring. If the spring length difference is marked, renew the springs as a pair.
4   The two oilseals fitted to each lower leg must be renewed if oil leakage has occurred. The seals may be prised from position, using a screwdriver. Refit the seals with the spring facing side inwards.
5   Cracked or perished dust excluders should be renewed as they prevent road dirt from reaching the oil seals and help prevent scoring of the fork stanchions.

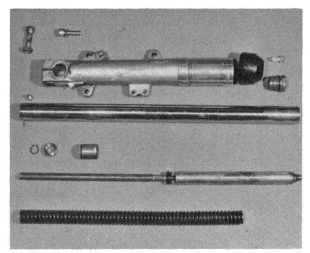

4.1 Front fork components - general view

4.4 Double oil seals can be prised out

## 5  Steering head bearings: examination and renewal

1   When the fork has been removed, the upper bearing inner race remains in the steering head and the lower inner race remains on the steering column. Both outer races remain in the steering head.
    The upper bearing inner race may be lifted out of the outer race. The lower bearing inner race must be pulled off the steering column but only if it requires renewal.
2   Clean and examine the outer bearing tracks whilst in the steering head. Since the forks rotate through only a small angle, the commonest damage to the bearings is brinelling. This is indenting of the roller tracks by the rollers, generally due to maladjustment. It can be felt, when turning the forks, by the steering seeming to 'index' in one position.
3   Outer races must be driven out using a soft metal drift. When refitting, clean the housings and make sure that the races seat squarely. Do not interchange parts of bearings.
4   Grease the bearings before reassembly. Fit the lower bearing inner race onto the steering column and place the upper bearing inner race into the steering head. Insert the steering column carefully into the steering head, holding the upper bearing in place. Continue reassembly in reverse order of dismantling.

## 6  Front forks: reassembly and replacement

1   It is essential to observe absolute cleanliness when reassembling the fork legs. Reassemble in the reverse order of that given for dismantling.

2   When replacing the spring seat and spring guides on the damper rod, ensure that the cutaways engage correctly with the ears on the rod end. Note the locating projection on the spring seat which must align with the recess in the inside of the fork leg. Before refitting the stanchion top bolts and the damper retaining screws, refill each fork leg with the specified quantity and grade of lubricant. Check that both drain plugs are tightened, before filling.

3   Before fully tightening the front wheel spindle clamps and the fork yoke pinch bolts, bounce the forks several times to ensure that they work freely and are clamped in their original settings. Complete the final tightening from the front wheel spindle clamps upwards. This will help align the fork legs correctly.

4   Before the machine is used on the road, check the adjustment of the steering head bearings. If they are too slack, judder will occur. There should be no detectable play in the head races when the handlebars are pulled and pushed, with the front brake applied hard.

5   Overtight head races are equally undesirable. It is possible to unwittingly apply a loading of several tons on the head bearings by overtightening, even though the handlebars appear to turn quite freely. Overtight bearings will cause the machine to roll at low speeds and give generally imprecise handling with a tendency to weave. Adjustment is correct if there is no perceptible play in the bearings and the handlebars will swing to full lock in either direction, when the machine is on the centre stand with the front wheel clear of the ground. Only a slight tap should cause the handlebars to swing.

## 7  Frame: examination

1   When the machine is stripped for overhaul, an excellent opportunity arises to inspect the frame for signs of cracks or damage. Look especially closely around the steering head and rear swinging arm pivots. Frame repairs must be entrusted to a specialist, who will have the equipment necessary to ensure correct alignment. In the event of damage, replacement is the only safe course.

2   If the front forks are removed, a quick visual alignment may be made by inserting a close-fitting tube in the steering head. When viewed from the front of the machine, the tube should align exactly with the centre line of the frame. Deviation will indicate damage to the front of the frame. Since the rear wheel of the Moto Guzzi does not have to be adjusted, wheel misalignment will indicate frame distortion. More accurate checking must be carried out with the frame stripped completely.

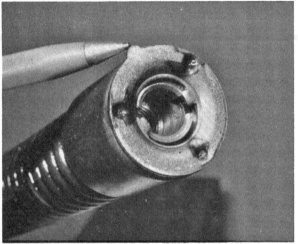

6.1a Note projection on the spring seat

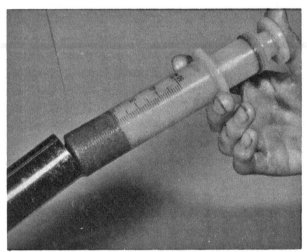

6.1b Do not omit to fill each fork leg with lubricant

## 8  Swinging arm: removal, examination and renovation

1   The rear fork assembly pivots on tapered roller bearings, fitted each side of the fork tubular cross-member. They bear on adjustable screw pivot stubs fitted into the lugs welded to the rear frame tube junctions.

2   When wear necessitates bearing renewal, rear suspension removal should be carried out as follows. Place the machine on the centre stand, raised on blocks so that the weight of the machine is biased towards the front wheel. Ensure that the machine is resting securely. If necessary, place a block below the front wheel to prevent the machine rolling forwards, off the centre stand.

3   Loosen the clamp which secures each silencer to the H section balance pipe. Unscrew the silencer mounting bolts and pull the silencers from position.

4   On V-1000 models, disconnect the parking brake cable at the operating lever on the caliper. Remove the caliper as a unit after detaching the two bolts. On all models with drum rear brakes, disconnect the brake rod by unscrewing the knurled adjuster nut and applying the rear brake pedal. Push out the trunnion from the operating arm to avoid loss.

5   On all models, loosen and remove the rear wheel spindle nut and then loosen the pinch bolt which passes through the left-hand arm of the rear fork. Withdraw the wheel spindle by passing a tommy bar through the hole in the spindle head. On disc rear brake models, move the caliper backwards slightly to clear the anchor lug on the swinging arm fork and then rotate the wheel backwards so that the caliper moves around the disc. When the caliper is free of the disc, move it forwards and tie it to a suitable portion of the frame. The hose clip on the fork member should be detached to free the cable.

6   Remove the wheel spacer from the centre of the wheel hub and then pull the rear wheel across to the left and off the drive splines. On all but 750S models, where the rear portion of the mudguard can be hinged up, it will be necessary to tilt the machine over to the right-hand side, to allow the wheel to clear the mudguard. An assistant should be enlisted to help during this operation.

7   Detach the lower end of the right-hand suspension unit from the bevel drive box after removing the nut and washer from the stud. Loosening of the upper mounting bolt may be necessary in order to give sufficient lateral movement of the damper. Support the weight of the drive box and remove the four securing nuts. Lift the bevel drive box away to the rear. Drainage of the lubricant is not required, provided that the bevel box is stored in an upright position with the input shaft pointing slightly upwards.

8   Remove the two socket screws on both sides of the machine which secure the footrest brackets, or on touring machines the silencer and rear crashbar mounting plate. It is necessary to remove these bolts and nuts only because they obstruct movement of the swinging arm cross-member during removal.

9   Detach the left-hand rear suspension unit from the swinging arm unit. Support the end of the swinging arm on a wooden block to prevent it dropping and fouling the frame lugs. Unscrew completely, and detach the two screw clips which secure the gaiter to the gearbox output boss and the swinging arm. Pull the rubber gaiter forwards off the swinging arm boss.

10  Loosen and remove the swinging arm pivot adjuster locknuts. These are usually very tight. The swinging arm, together with the final driveshaft, may be withdrawn towards the rear after unscrewing the two pivot stubs. Tilt the assembly to clear the frame lugs.

11  Remove the final driveshaft and universal joint from the swinging arm. The lower boss of the universal joint should be a tight

push fit in the ball bearing contained within the swinging arm fork. Use a wooden drift to remove the joint. If the joint boss is loose in the bearing inner race, wear has developed. See Section 9 of this Chapter for further details.

12  Remove the spacer collar from one side of the swinging arm cross-member and prise out the oil seal with a screwdriver. Lift the tapered bearing cage/inner race from position. Repeat for the other bearing assembly. Do not allow the bearings to become interchanged.

13  Wash the bearings thoroughly in petrol before examination is carried out. These bearings have a very long life provided lubrication is not neglected. Slight wear may be taken up by adjustment of the pivot stubs. Check for looseness of the rollers and for pitting or scoring of the roller tracks. If the bearings require renewal, the outer races must be pulled from position, using a special extractor. Neither race can be drifted from place, access being impossible due to the crossmember design. Many motorcycle repair specialists will have a suitable expanding puller and will be willing to carry out this operation for a nominal sum.

14  The swinging arm fork may be refitted by reversing the dismantling procedure. Ensure that the bearings are greased thoroughly before refitting the seals as no provision is given for subsequent lubrication. Adjustment of the tapered roller bearings must be made before the bevel drive box or suspension units are refitted. Screw in each pivot stub until the swinging arm fork is central in the frame. Check this by measuring the distance the pivot stubs protrude outwards, using a vernier gauge or ruler. Screw the pivot stubs further inwards an equal amount until a small amount of resistance in the bearings can be felt when raising and lowering the swinging arm fork. At this point there should be no perceptible side-to-side play. Fit and tighten the locknuts without allowing the pivot stubs to turn.

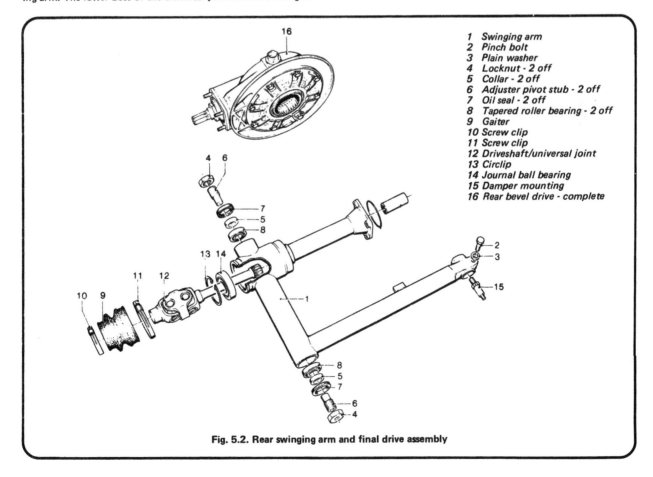

1   Swinging arm
2   Pinch bolt
3   Plain washer
4   Locknut - 2 off
5   Collar - 2 off
6   Adjuster pivot stub - 2 off
7   Oil seal - 2 off
8   Tapered roller bearing - 2 off
9   Gaiter
10  Screw clip
11  Screw clip
12  Driveshaft/universal joint
13  Circlip
14  Journal ball bearing
15  Damper mounting
16  Rear bevel drive - complete

**Fig. 5.2. Rear swinging arm and final drive assembly**

8.3 Loosen the silencer clamps and remove the single mounting bolt

8.5a Remove the rear wheel nut and washer ...

8.5b ... loosen the spindle clamp bolt and ...

8.5c ... draw the spindle out from the left

8.5d Pull the caliper back off the brake disc

8.7a The rear suspension dampers are retained by a bolt and ...

8.7b ... by a nut on a stud. Note split rubber bushes

8.7c Detach the bevel drive box and ...

8.7d ... hook out the shaft splined sleeve

8.9 Pull back the rubber gaiter at the gearbox joint

8.10a Unscrew the pivot adjuster locknut and ...

8.10b ... remove both adjuster pivot stubs

8.13 Check the outer races for indentation and scoring

8.14a Grease the tapered bearing thoroughly

8.14b Refit the oil seals followed by ...

8.14c ... the spacer collars

## 9  Final driveshaft: examination and renovation

1   The final driveshaft may be removed for inspection after
detaching the swinging arm and bevel drive box, as described in
the previous Section.
2   If wear in the universal joint is evident, the complete
assembly must be renewed. On all but V-1000 models the driveshaft is a separate item to the joint and need only be renewed if
the coupling splines at either end are badly worn
3   Check the fit of the driveshaft ball bearing inner race with
the inner boss on the universal joint. The boss should be a tight
drive fit in the race. Looseness at this point is not an uncommon
occurrence. The only remedy is to renew one or both of the
components. The use of a locking fluid is unlikely to provide a
reliable remedy unless the clearance is very small. The bearing
is retained within the swinging arm fork by a large internal
circlip. Removal of the circlip may prove difficult unless a pair
of long handled circlip pliers is available. The bearing may be
drifted out, using a long punch inserted from the bevel drive end.
4   When refitting the driveshaft, grease the internal and external
splines of the coupling sleeve and shaft with graphite grease.

9.1a One piece final drive shaft - V-1000 models

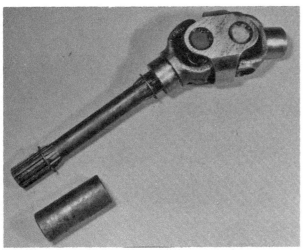

9.1b Two piece final drive shaft - All 5-speed models

9.3 Shaft bearing is retained by a circlip

**10 Rear bevel drive - examination and renovation**

1   Dismantling the bevel drive is beyond the scope of this book and the majority of amateur mechanics. Wear or damage will be indicated by a high pitched whine. 'Backlash' between the crownwheel and pinion may be assessed by holding the output shaft firmly and rotating the input shaft in both directions. Any lateral play in the crownwheel can be felt by pulling and pushing the output driveshaft.

2   A faulty output shaft oil seal will be indicated by excessive oil in the brake drum, or around the wheel spindle nut.

3   The helical drive coupling and rear wheel drive splines should be examined for wear or damage.

**11 Rear suspension units: adjustment, removal and examination**

1   The hydraulically-damped rear suspension units may be adjusted to suit the road conditions and the load carried. On all but the 750, V-1000 and Le Mans models, the units are adjustable to five positions by means of a lever integral with the suspension unit. On Le Mans, V-1000 and 750 machines, the units are adjusted by means of a special 'C' spanner supplied with the tool kit. Three different load settings are possible in these latter cases. Both units **must** be set to the same position. Faulty dampers may be suspected if handling becomes unpredictable, or if the rear end of the machine bounces up and down when depressed and released. Oil leaks will indicate a faulty oil seal.

2   The damper units fitted to V-1000, Le Mans and 750 models, should be returned to the manufacturers for overhaul as their dismantling requires the use of special tools. The five position dampers can, however, be partially dismantled for inspection. Place one unit in position No. 1 and compress the spring to allow removal of the upper spring seat. Two people should carry out this operation because the springs are heavy. Remove the spring and adjuster cam.

3   Check damper action by pulling and pushing on the damper rod. It should require more effort to pull it out than to push it in. In both directions, the action must be uniform over the total length of the stroke. If oil is escaping, the damper rod seal is faulty. It is not possible to renovate the dampers, they can only be renewed.

4   Check the spring free length (see Specifications). Stronger springs are available for use when heavy loads are habitually carried.

5   The conical rubber bushes which support the damper mounting eyes should be renewed if they have perished or become compacted.

**12 Centre stand: examination**

1   The centre stand pivots on two short bolts passing through lugs on the lower frame tubes. Each bolt is fitted with a separate bush. Two return springs are provided, each of which is anchored to a drilled plate which also serves as a stop when the stand is in the extended position.

2   Check the pivot bolts for security. Occasionally, remove the bolts and lubricate the bushes to prevent premature wear. If the spring becomes weakened or the hook ends have worn, it should be renewed.

11.1 Rear suspension units are adjustable for ride

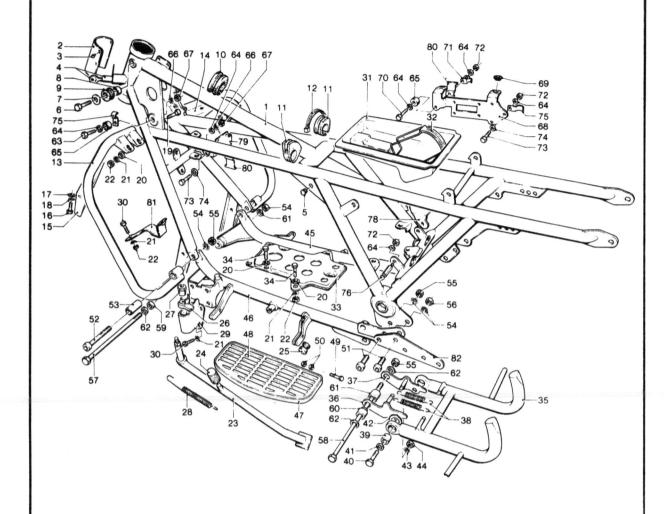

**Fig. 5.3. Frame assembly**

| | | | | | | |
|---|---|---|---|---|---|---|
| 1 | Frame | 21 | Star washer - 7 off | 41 | Washer - 2 off | 62 | Washer - 3 off |
| 2 | Frame number plate | 22 | Nut - 7 off | 42 | Star washer - 2 off | 63 | Bolt - 2 off |
| 3 | Rivet - 2 off | 23 | Prop stand arm | 43 | Star washer - 2 off | 64 | Washer - 7 off |
| 4 | Steering lock | 24 | Rubber | 44 | Nut - 2 off | 65 | Rubber sleeve - 2 off |
| 5 | Bung - 4 off | 25 | Buffer | 45 | RH sub frame member | 66 | Star washer - 6 off |
| 6 | Bolt - 2 off | 26 | Push plate | 46 | LH sub frame member | 67 | Nut - 4 off |
| 7 | Plain washer - 2 off | 27 | Sleeve nut | 47 | Footboard - 2 off | 68 | Rectifier bracket |
| 8 | Sleeve - 2 off | 28 | Return spring | 48 | Rubber inlay - 2 off | 69 | Rubber cushion - 2 off |
| 9 | Rubber sleeve - 2 off | 29 | Stand mounting bracket | 49 | Pivot bolt - 4 off | 70 | Bolt - 2 off |
| 10 | Tank seat | 30 | Bolt - 6 off | 50 | Nut - 8 off | 71 | Clip |
| 11 | Tank seat - 2 off | 31 | Tool tray | 51 | Socket screw - 4 off | 72 | Nut - 3 off |
| 12 | Strap - 3 off | 32 | Strap | 52 | Socket screw - 2 off | 73 | Bolt - 2 off |
| 13 | Crash bar - complete | 33 | Battery mounting plate | 53 | Spacer - 2 off | 74 | Washer |
| 14 | RH crash bar | 34 | Bolt - 7 off | 54 | Star washer - 6 off | 75 | Clip - 2 off |
| 15 | LH crash bar | 35 | Centre stand | 55 | Nut - 8 off | 76 | Retaining pin - 2 off |
| 16 | Bolt - 4 off | 36 | LH spring plate | 56 | Nut | 77 | |
| 17 | Washer - 9 off | 37 | RH spring plate | 57 | Front engine bolt | 78 | Bracket |
| 18 | Star washer - 4 off | 38 | Return spring - 2 off | 58 | Rear gearbox bolt | 79 | Coil bracket |
| 19 | Bolt - 2 off | 39 | Bush - 2 off | 59 | Spacer | 80 | Clip |
| 20 | Washer - 9 off | 40 | Bolt - 2 off | 60 | Spacer - 2 off | 81 | Bracket - 2 off |
| | | | | 61 | Spacer - 2 off | 82 | |

12.1 Arrangement of centre stand mountings

### 13 Prop stand: examination

1   The prop stand fitted to all but the touring models is of a
conventional type, retained at the front left-hand side of the
machine by the engine forward mounting bolt. 850 California
models employ a side stand, pivoting from a separate bracket in
front of the left-hand footboard. A locking device prevents the
stand retracting when the weight of the machine is on it. The
V-1000 prop stand is similar to that of the 850 California model
but incorporates a cable pull mechanism which operates the
caliper parking brake.
2   Check that the stand extends and retracts correctly,
lubricating the pivot where necessary with oil or grease. Ensure
that the extension spring is in good condition. A stand which
falls when the machine is being ridden may have disasterous
results when cornering, almost certainly unseating the rider.
3   Refer to Chapter 6, Section 16 for details of the V-1000
stand-operated parking brake.

### 14 Footrests and foortboards

1   On touring models, footboards are provided for the rider in
place of the more normal footrests. Each footboard is mounted
on two pivot bolts, retained by a nut and locknut. By this means
the footboards may be adjusted to fold upwards easily and so
present less resistance when grounding is experienced in cornering. The passenger foot rests are incorporated in the rear crashbar, which has footrest rubbers placed over the lower horizontal
portion. To replace the rubbers, undo the lower and upper
mounting bolts of each crashbar and pull the rubbers from
place. Soap or French Chalk may be used to aid fitting of new
rubbers onto the bars.
2   All other models have footrests mounted on a forged bar,
one of which is secured to each side of the machine by the bolts
retaining the rear of the engine sub-frame assembly. The rear
footrests are hinged on the bars to allow folding when not in
use. The footrest rubbers may be renewed as individual
components as can the folding portion of the rear footrests. The
hinge is formed by a simple nut and bolt.

### 15 Dualseat: removal

1   Lift the dualseat and remove the two bolts which serve as
pivot pins. The seat can then be lifted away from the machine.
2   Before refitting the seat, lubricate the pivots and also the
mechanism which locks the seat when down.

### 16 Steering head lock

1   The only maintenance possible for the lock is to lubricate
the barrel occasionally - not the keyhole. Use a light universal
oil.
2   If the key is lost it is possible to obtain a replacement,
provided the key number is known. If no number is available
or the lock malfunctions the lock securing dowel must be drilled
out and a new lock fitted.

### 17 Instrument drive cables: examination and replacement

1   Drive cables should be examined and lubricated occasionally.
The outer sheath should be examined for cracks or damage, the
inner cable for broken or frayed strands. Jerky or sluggish
instrument movement is generally caused by a faulty cable.
2   Detach the cable at the drive end, and withdraw the inner
cable. Clean and examine the cable. Re-lubricate it with high
melting point grease, but do not grease the top six inches of
cable, at the instrument end, or grease will work its way into the
instrument head and ruin the movement.
3   Route the cables as they were originally. Make sure that
the steering turns freely.

### 18 Instruments: removal

1   Before removal of the instruments is undertaken disconnect
the battery to prevent inadvertent shorting of any of the circuits
involved. Detach the speedometer drive cable and where fitted
the tachometer drive cable. Both are retained by knurled gland
nuts.
2   Detach the bulb holders and bulbs and the electrical
connections, where necessary, after removing the base of the
lighting console. It is held by four knurled rings on all models
except on the V-1000 model, which utilises four screws.
3   Remove the two bolts securing the instrument console to the
front forks and lift the assembly away. Note the sequence of
rubber bushes, sleeves and washers.
4   The instruments cannot be repaired. Remember that it is
necessary to have a functioning speedometer accurate within $\pm$
10% at 30 mph, in the UK. If the odometer continues to record
when the speedometer fails to function, the instrument head is
faulty and requires renewal or repair.

### 19 Cleaning the machine

1   After removing all surface dirt with a rag or sponge which is
washed frequently in clean water, the machine should be allowed
to dry thoroughly. Application of car polish or wax to the cycle
parts will give a good finish, particularly if the machine receives
this attention at regular intervals.
2   The plated parts should require only a wipe with a damp rag,
but if they are badly corroded, as may occur during the winter
when the roads are salted, it is permissible to use one of the
proprietary chrome cleaners. These often have an oily base which
will help to prevent corrosion from recurring.

3   If the engine parts are particularly oily, use a cleaning compound such as Gunk or Jizer. Apply the compound whilst the
parts are dry and work it in with a brush so that it has an opportunity to penetrate and soak into the film of oil and grease.
Finish off by washing down liberally, taking care that water
does not enter the carburettors, air cleaners or the electrics. If
desired, the now clean aluminium alloy parts can be enhanced
still further when they are dry by using a special polish such as

Solvol Autosol. This will restore the full lustre.
4   If possible, the machine should be wiped down immediately
after it has been used in the wet, so that it is not garaged under
damp conditions which will promote rusting. Remember there is
less chance of water entering the control cables and causing
stiffness if they are lubricated regularly as described in the
Routine Maintenance Section.

## 20 Fault diagnosis: frame and forks

| Symptom | Cause | Remedy |
|---|---|---|
| Machine veers either to the left or the right with hands off handlebars | Bent frame<br>Twisted forks | Check and renew.<br>Check and renew if necessary. |
| Machine rolls at low speed | Overtight steering head bearings | Slacken until adjustment is correct. |
| Machine judders when front brake is applied | Slack steering head bearings | Tighten until adjustment is correct. |
| Machine pitches on uneven surfaces | Ineffective fork dampers<br>Ineffective rear suspension units<br>Suspension too soft | Renew sealed units.<br>Check whether units still have damping action.<br>Raise suspension unit adjustment one notch. |
| Fork action stiff | Fork legs out of alignment (twisted in yokes) | Slacken yoke clamps, and fork top bolts.<br>Pump fork several times then retighten from bottom upwards. |
| Machine wanders. Steering imprecise Rear wheel tends to hop | Worn swinging arm pivot | Dismantle and renew bearings and pivot shaft. |

# Chapter 6 Wheels, brakes and tyres

## Contents

## Specifications

### Tyres

| Sizes | | Front | Rear |
|---|---|---|---|
| 750S and S3 ... ... ... ... ... ... ... ... | | 3.25 H 18 | 3.50 H 18 |
| 850T and T3 ... ... ... ... ... ... ... ... | | 3.50 H 18 | 4.10 H 18 |
| 850 Le Mans ... ... ... ... ... ... ... ... | | 3.50 H 18 | 4.10 V 18 |
| V-1000 ... ... ... ... ... ... ... ... ... | | 4.10 H 18 | 4.10 H 18 |

all tyres 18 inch diameter.
*Tyres rated as 'H' are safe up to speeds of 130 mph. 'V' rated tyres are safe at speeds above 130 mph.

### Tyre pressures

| | Front | Rear |
|---|---|---|
| 750S, S3, 850T3 and Le Mans ... ... ... ... ... ... | 29 psi (2 kg cm$^2$) | 33 psi (2.3 kg cm$^2$) |
| 850T ... ... ... ... ... ... ... ... ... | 26 psi (1.8 kg cm$^2$) | 33 psi (2.3 kg cm$^2$) |
| V-1000 ... ... ... ... ... ... ... ... ... | 30 psi (2.1 kg cm$^2$) | 34 psi (2.4 kg cm$^2$) |

*When carrying a pillion passenger the rear tyre pressure should be increased by 3 psi (0.2 kg cm$^2$).
When travelling at continuous high speeds an additional 3 psi (0.2 kg cm$^2$) should be added to both front and rear tyres.

### Brakes

| | Front | Rear |
|---|---|---|
| 750S ... ... ... ... ... ... ... ... ... | Twin 300 mm (11.8 in) discs | Single leading shoe drum brake 220 x 25 mm (8.66 x 0.98 in) |
| 850T ... ... ... ... ... ... ... ... ... | Single 300 mm (11.8 in) disc | As above |
| All other ... ... ... ... ... ... ... ... ... | Twin 300 mm (11.8 in) disc | Single 242 mm (9.5 in) disc |

## 1 General description

All models are fitted with 18 inch diameter wheels, both at the front and rear. With the exception of the Le Mans model, which has one-piece cast aluminium alloy wheels, all models are fitted with aluminium safety rims laced to cast aluminium hubs by chrome finished spokes.

The front brake of the 750S and 850T models is an hydraulically operated disc unit mounted on the right-hand side of the hub and operated by a handlebar lever. The rear brake is a rod operated drum unit. All other machines are fitted with twin front disc brakes and a single rear disc brake. The right-hand front disc is operated by the normal method. The rear disc and left-hand front disc are interconnected and operated by the rear brake pedal. The integration of the brakes in this way, using a system patented by Moto Guzzi, allows the rear brake pedal only to be used during normal riding. A compensator unit in the system controls the braking force applied so that the front brake receives about 75% of the effort and the rear 25%.

## 2  Front wheel: examination and renovation (spoked wheel models)

1   Place the machine on the centre stand so that the front wheel is raised clear of the ground. Spin the wheel and check the rim alignment. Small irregularities can be corrected by tightening the spokes in the affected area, although a certain amount of practice is necessary to prevent over-correction. Any flats in the wheel rim should be evident at the same time. These are more difficult to remove and in most cases it will be necessary to have the wheel rebuilt on a new rim. Apart from the effect on stability, a flat will expose the tyre bead and walls to greater risk of damage.

2   Check for loose or broken spokes. Tapping the spokes is the best guide to tension. A loose spoke will produce a quite different sound and should be tightened by turning the nipple in an anticlockwise direction. Always recheck for run-out by spinning the wheel again.  If the spokes have to be tightened an excessive amount, it is advisable to remove the tyre and tube by the procedure detailed in Section 17 of this Chapter; this is so that the protruding ends of the spokes can be ground off, to prevent them from chafing the inner tube and causing punctures.

## 3  Front wheel: examination and renovation (cast alloy wheel models)

1   Carefully check the complete wheel for cracks and chipping,

particularly at the spoke roots and the edge of the rim. As a general rule a damaged wheel must be renewed as cracks will cause stress points which may lead to sudden failure under heavy load. Small nicks may be radiused carefully with a fine file and emery paper (No. 600 - No. 1000) to relieve the stress. If there is any doubt as to the condition of a wheel, advice should be sought from a Moto Guzzi repair specialist.

2   Each wheel is covered with a coating of lacquer, to prevent corrosion. If damage occurs to the wheel and the lacquer finish is penetrated, the bared aluminium alloy will soon start to corrode. A whitish grey oxide will form over the damaged area, which in itself is a protective coating. This deposit however, should be removed carefully as soon as possible and a new protective coating of lacquer applied.

3   Check the lateral run-out at the rim by spinning the wheel and placing a fixed pointer close to the rim edge. If the maximum run-out is greater than 1.0 mm (0.040 in), it is recommended that the wheel be renewed. This is, however, a counsel of perfection; a run-out somewhat greater than this can probably be accommodated without noticeable effect on steering. No means is available for straightening a warped wheel without resorting to the expense of having the wheel skimmed on all faces. If warpage was caused by impact during an accident, the safest measure is to renew the complete wheel. Worn wheel bearings may cause rim run-out. These should be renewed as described in Section 10 of this Chapter.

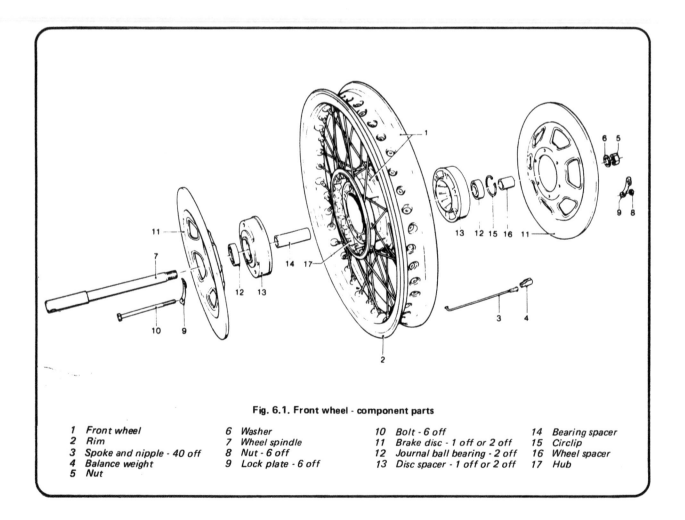

Fig. 6.1. Front wheel - component parts

| | | |
|---|---|---|
| 1  Front wheel | 6  Washer | 10  Bolt - 6 off | 14  Bearing spacer |
| 2  Rim | 7  Wheel spindle | 11  Brake disc - 1 off or 2 off | 15  Circlip |
| 3  Spoke and nipple - 40 off | 8  Nut - 6 off | 12  Journal ball bearing - 2 off | 16  Wheel spacer |
| 4  Balance weight | 9  Lock plate - 6 off | 13  Disc spacer - 1 off or 2 off | 17  Hub |
| 5  Nut | | | |

## 4  Front disc brake: examination and renovation

1   Check the front brake master cylinder, hoses and caliper units for signs of leakage. Pay particular attention to the condition of the hoses, which should be renewed without question if there are signs of cracking, splitting or other exterior damage. On machines with integrated braking the front left-hand disc is operated by the rear brake hydraulic circuit. For the purposes of maintenance (except for fluid level inspection) the procedures are identical.

2   Check the level of hydraulic fluid by removing the cap on the brake fluid reservoir and lifting out the diaphragm plate. The condition of the fluid can be checked at the same time. Checking the fluid level is one of the maintenance tasks which should never be neglected. If the fluid is below the lower level mark, brake fluid of the correct specification must be added. **Never** use engine oil or any fluid other than that recommended. Other fluids have unsatisfactory characteristics and will rapidly destroy the seals.

3   The sets of brake pads should be inspected for wear after prising off the corrugated inspection plate from the top of each caliper. If either pad is less than 6 mm (0.2362 in) thick, both pads must be renewed as a set. Where twin disc brakes are utilised it is likely that the right-hand set of pads will wear less quickly; since the handlebar lever operated brake is used relatively rarely.

4   The pads may be removed with the wheel in place. Depress the pin keeper spring at one end and withdraw one pin. Remove the second long pin and the spring. The central taper pin can now be removed followed by the two pads, one at a time.

5   The pads set on the integrated front brake can wear very quickly. Where a 'riding on the brakes' style of riding is used, the pad life may be as little as 2000 miles.

## 5  Front brake caliper: overhaul

1   Caliper removal and replacement can take place without removing the front wheel. Where twin disc brakes are utilised,

each caliper should be removed and dismantled individually, using an identical procedure. Again it should be remembered that the left-hand caliper is operated by the rear brake hydraulic circuit.

2   Detach the brake feed pipe from the caliper, allowing any fluid to drain into a suitable container. **Do not** allow any fluid to contact the paintwork. It is a superb paintstripper. Remove the two bolts which secure the caliper unit to the fork leg, and lift the complete caliper from position on the disc.

3   Prise off the inspection cover and remove the brake pads, as described in the previous Section. The caliper consists of two cylinders and pistons retained by two bolts and interconnected by a passageway that allows the brake fluid pressure to equalise. Remove the two securing bolts and separate the two components. Note the O ring which seals the passageway.

4   Carefully prise off the dust excluding boot from one piston. The piston can be displaced most easily using an air hose or tyre pump connected to the fluid feed orifice. Using a fine pointed instrument dislodge the sealing O ring from the groove in the cylinder. Dismantle the other caliper half in a similar manner.

5   Wash the brake caliper components in clean hydraulic fluid. **DO NOT** use petrol or other solvents to clean brake parts as the rubber components will be damaged. Inspect the pistons, and cylinder bores for scoring which may lead to leakage. If damage is evident, renew the components affected.

6   Reassemble the caliper by reversing the dismantling procedure. Owing to their low cost it is recommended that all seals and boots be renewed without question. When refitting the O ring seals, apply a little hydraulic fluid as lubricant. Special brake component grease which has a very high melting temperature should be used to lubricate the pistons. If required, the caliper may be centralised on the disc using shims placed between the caliper and the mounting lug on the fork leg.

7   Note that work on hydraulic brake systems must be carried out in scrupulously clean conditions. Particles of dirt will score the working parts and cause early failure of the system.

8   After reassembling and refitting the caliper units, the hydraulic systems must be bled of all air as described in Section 7 of this Chapter.

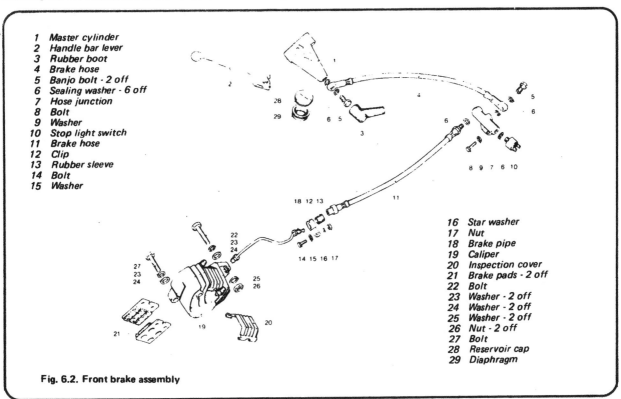

1   Master cylinder
2   Handle bar lever
3   Rubber boot
4   Brake hose
5   Banjo bolt - 2 off
6   Sealing washer - 6 off
7   Hose junction
8   Bolt
9   Washer
10   Stop light switch
11   Brake hose
12   Clip
13   Rubber sleeve
14   Bolt
15   Washer

16   Star washer
17   Nut
18   Brake pipe
19   Caliper
20   Inspection cover
21   Brake pads - 2 off
22   Bolt
23   Washer - 2 off
24   Washer - 2 off
25   Washer - 2 off
26   Nut - 2 off
27   Bolt
28   Reservoir cap
29   Diaphragm

**Fig. 6.2. Front brake assembly**

4.3 Prise off the cover to gain access to the pads

4.4a Depress spring to remove pins

4.4b Pads will lift out individually

5.2 Each caliper is retained by two socket screws

5.6a Check centralisation of disc between caliper halves

5.6b Fit shims as required to centralise caliper

### 6   Front disc brake master cylinder: examination and renovation (right-hand brake only)

1   The master cylinder and hydraulic reservoir take the form of a combined unit mounted on the right-hand side of the handlebars, to which the front brake lever is attached. The master cylinder is actuated by the front brake lever, and applies hydraulic pressure through the system to operate the front brake when the handlebar lever is manipulated. The master cylinder pressurises the hydraulic fluid in the brake pipe which, being incompressible, causes the piston to move in the caliper unit and apply the friction pads to the brake. If the master cylinder seals leak, hydraulic pressure will be lost and the braking action rendered much less effective.

2   Before the master cylinder can be removed the system must be drained. Place a clean container below the caliper unit and attach a plastic tube from one bleed screw on top of the caliper unit to the container. Open the bleed screw one complete turn and drain the system by operating the brake lever until the master cylinder reservoir is empty. Close the bleed screw and remove the pipe.

3   Place a rag under the banjo union connecting the brake hose to the master cylinder. Pull back the rubber boot and remove the banjo bolt. Loosen the throttle twist grip and slide it off the handlebar end. Slacken the pinch bolt securing the master cylinder/handlebar lever and remove the complete assembly.

4   Remove the reservoir cap and lift out the rubber diaphragm. Detach the operating lever from the master cylinder by unscrewing the nut and bolt. To remove the master cylinder piston, select a short rod that can be inserted through the fluid outlet orifice. The rod should have a smooth rounded end, to prevent damage to the master cylinder interior. Gently tap the rod and drive out the piston and locking ring.

5   From the master cylinder, remove the piston return spring and the spring guide. Pull the lockring off the piston, followed by the thin washer and scraper ring. Check the condition of the two piston rings. If they are unmarked and no leakage was evident, they may be reused.

6   The component parts of the master cylinder assembly and the caliper assembly may wear or deteriorate in function over a long period of use. It is however, generally difficult to foresee how long each component will work with proper efficiency and from a safety point of view it is best to change all the expendable parts every two years on a machine that has covered a normal mileage.

7   Replace the master cylinder by reversing the dismantling procedure. After reconnecting the brake hose, the system must be bled as described in the next Section. Free play at the handlebar lever should be adjusted so that there is 0.15 mm (0.006 in) between the lever and the piston face. Adjustment is effected by a grubscrew fitted to the lever.

8   When positioning the master cylinder, place it so that it is as upright as possible, yet in such a position that the lever operating angle is compatible with easy operation.

### 7   Bleeding the hydraulic brake system

#### Right-hand front disc brake

1   If the hydraulic system has to be drained and refilled, if the front brake lever travel becomes excessive or the lever operates with a soft or spongy feeling, the brakes must be bled to expel air from the system. The procedure for bleeding the hydraulic brake is best carried out by two persons.

2   First check the fluid level in the reservoir and top up with fresh fluid.

3   Keep the reservoir at least half full of fluid during the bleeding procedure.

4   Screw the cap on to the reservoir to prevent a spout of fluid or the entry of dust into the system. Place a clean glass jar below the caliper bleed screw and attach a clear plastic pipe from both caliper bleed screws to the container. Place some clean hydraulic fluid in the jar so that the pipes are always immersed below the surface of the fluid.

5   Unscrew the bleed screws one half turn and squeeze the brake lever as far as it will go but do not release it until the bleeder valves are closed again. Repeat the operation a few times until no more air bubbles come from the plastic tube.

6   Keep topping up the reservoir with new fluid. When all the bubbles disappear, close the bleeder screws securely. Remove the plastic tubes and install the bleeder valve dust caps. Check the fluid level in the reservoir, after the bleeding operation has been completed.

7   Reinstall the diaphragm and tighten the reservoir cap securely. Do not use the brake fluid drained from the system, since it will contain minute air bubbles.

8   Never use any fluid other than that recommended. Oil must not be used under any circumstances.

#### Integrated braking circuit

9   Bleed the brakes in a manner similar to that given for the single front disc brake circuit. Bleed the front caliper first followed by the rear caliper. If the brake action is still spongy, bleed the front caliper again.

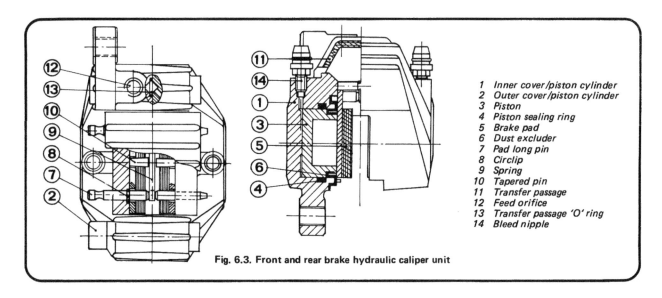

1   Inner cover/piston cylinder
2   Outer cover/piston cylinder
3   Piston
4   Piston sealing ring
5   Brake pad
6   Dust excluder
7   Pad long pin
8   Circlip
9   Spring
10   Tapered pin
11   Transfer passage
12   Feed orifice
13   Transfer passage 'O' ring
14   Bleed nipple

Fig. 6.3. Front and rear brake hydraulic caliper unit

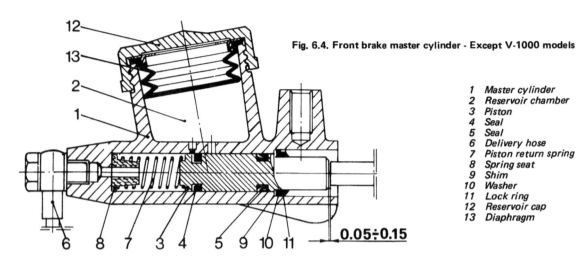

**Fig. 6.4. Front brake master cylinder - Except V-1000 models**

1   Master cylinder
2   Reservoir chamber
3   Piston
4   Seal
5   Seal
6   Delivery hose
7   Piston return spring
8   Spring seat
9   Shim
10  Washer
11  Lock ring
12  Reservoir cap
13  Diaphragm

0.05÷0.15

7.4 Connect pipes to both nipples when bleeding brakes

### 8   Integrated brake system and rear caliper: maintenance and overhaul

1   The rear brake caliper is identical to the right-hand brake caliper. The checking of the pads, their overhaul and general maintenance is therefore the same.
2   Refer to Sections 4 and 5 of this Chapter for the relevant details.

### 9   Integrated brake master cylinder: examination and renovation

1   The combined rear brake and front left-hand brake master cylinder is mounted on the right-hand side of the machine and is operated from the rear brake pedal via a short link rod.
2   Before the master cylinder is disconnected, the system should be drained of fluid. Place a plastic tube on the bleed screw of each caliper and open the screws one complete turn. Depress and release the brake pedal until all the fluid has been expelled.
3   Place a rag under the banjo union at the rear of the master cylinder. Remove the banjo bolt and allow any excess fluid to drip onto the rag. Disconnect the brake link rod from the bell crank on the master cylinder by removing the split pin and washer. On V-1000 models, detach the two wires that lead to the brake fluid level switch in the reservoir cap. Both wires are a push fit.

4   Remove the nut and washer from the rear of the bell crank pivot stub behind the frame lug. The master cylinder may be lifted away after removal of the second mounting, which is a socket bolt.
5   The master cylinder is fundamentally the same type of unit as that utilised for the single front brake. Refer to Section 6 of this Chapter for dismantling and overhaul details.
6   After refitting the master cylinder and bleeding the system, adjust the clearance between the end of the bell crank and the piston to 0.15 mm (0.006 in) by means of the adjuster screw forward of the crank pivot.

### 10   Front wheel bearings: examination and replacement

1   If wear has developed in the wheel bearings the front wheel should be removed to gain access to the bearings for renewal. Bearings should also be removed at the specified routine maintenance interval for cleaning, inspection and regreasing.
2   After the removal of the retaining circlip from the right-hand side of the hub the wheel bearings can be drifted out of position, using a suitable drift. Support the wheel so the exit of the bearing is not obstructed. When the first bearing has been removed the spacer that lies between the two bearings can be removed. Insert the drift and drive out the opposite bearing.
3   Remove all the old grease from bearings and hub. Wash the bearings in petrol and dry them thoroughly. Check the bearings for roughness by spinning them whilst holding the inner track with one hand and rotating the outer track with the other. If there is the slightest sign of roughness renew them.
4   Before driving bearings back into the hub, pack the hub with new grease and also grease the bearings. Use the same double diameter drift to place them in position.
5   Some machines are fitted with pre-packed ball bearings which are sealed on both sides. The lubricant with which the bearings are packed on assembly will last the life of the bearing.

### 11   Removing and replacing the discs

1   It is unlikely that the brake discs will require attention unless bad scoring has developed or the discs have warped. To detach the discs first remove the wheel in question.
2   Each disc is retained against a thick spacer on the hub by six long bolts passing through the hub, the nuts of which are secured in pairs by locking plates. Where two discs are employed on one hub the bolts pass through all the components. Bend down the ears of the locking plates and remove the nuts. Remove the disc complete with spacer and drift the spacer boss out of the centre of the disc.

## Tyre changing sequence - tubed tyres

 Deflate tyre. After pushing tyre beads away from rim flanges push tyre bead into well of rim at point opposite valve. Insert tyre lever adjacent to valve and work bead over edge of rim.

 Use two levers to work bead over edge of rim. Note use of rim protectors

Remove inner tube from tyre

When first bead is clear, remove tyre as shown

When fitting, partially inflate inner tube and insert in tyre

Work first bead over rim and feed valve through hole in rim. Partially screw on retaining nut to hold valve in place.

Check that inner tube is positioned correctly and work second bead over rim using tyre levers. Start at a point opposite valve.

Work final area of bead over rim whilst pushing valve inwards to ensure that inner tube is not trapped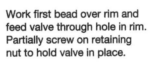

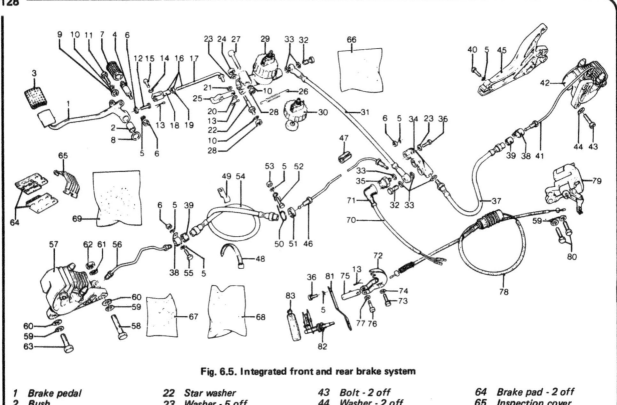

**Fig. 6.5. Integrated front and rear brake system**

| | | | | | | |
|---|---|---|---|---|---|---|
| 1 | Brake pedal | 22 | Star washer | 43 | Bolt - 2 off | 64 | Brake pad - 2 off |
| 2 | Bush | 23 | Washer - 5 off | 44 | Washer - 2 off | 65 | Inspection cover |
| 3 | Rubber | 24 | Self locking nut | 45 | Caliper mounting bracket | 66 | Master cylinder repair kit |
| 4 | Pivot pin | 25 | Bell crank | 46 | Brake pipe | 67 | Screw replacement kit |
| 5 | Star washer - 7 off | 26 | Return spring | 47 | Rubber sleeve | 68 | Clevis pin replacement kit |
| 6 | Nut - 5 off | 27 | Bolt | 48 | Strap | 69 | Seal kit |
| 7 | Rubber | 28 | Nut | 49 | Clip | 70 | Stop light lead |
| 8 | Washer | 29 | Master cylinder | 50 | Guide clip | 71 | Rubber boot |
| 9 | Washer | 30 | Cap/fluid level switch | 51 | Grommet | 72 | Parking brake lever |
| 10 | Star washer - 2 off | 31 | Brake hose | 52 | Bolt | 73 | Bolt |
| 11 | Bolt | 32 | Banjo bolt - 2 off | 53 | Nut | 74 | Washer |
| 12 | Bolt | 33 | Sealing washer - 6 off | 54 | Brake hose | 75 | Push bar |
| 13 | Split pin - 3 off | 34 | Hose junction | 55 | Bolt | 76 | Clevis pin |
| 14 | Washer | 35 | Stop light switch | 56 | Brake pipe | 77 | Washer - 2 off |
| 15 | Clevis pin | 36 | Bolt - 4 off | 57 | LH front brake caliper | 78 | Parking brake cable |
| 16 | Operating rod | 37 | Brake hose | 58 | Bolt | 79 | Parking brake caliper |
| 17 | Link rod | 38 | Clip - 2 off | 59 | Washer - 4 off | 80 | Bolt - 2 off |
| 18 | Clevis fork | 39 | Rubber sleeve - 2 off | 60 | Shim - A/R | 81 | Switch bracket |
| 19 | Locknut | 40 | Bolt | 61 | Star washer | 82 | Ignition cut-out switch |
| 20 | Washer | 41 | Brake pipe | 62 | Nut | 83 | Rubber boot |
| 21 | Star washer | 42 | Caliper | 63 | Bolt | | |

10.2a Right-hand wheel bearing is secured by circlip

10.2b Bearings may be drifted out

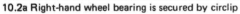

10.2c Note the bearing spacer; it must not be omitted

11.2 Disc is retained by 6 bolts and nuts

### 12 Rear wheel: examination and renovation

1   Place the machine on the centre stand so that the rear wheel is raised clear of the ground. Check for rim alignment, loose spokes etc, as described for the front wheel in Section 2 or 3, depending on the wheel type.

### 13 Rear drum brake: removal, examination and renovation. 750S and 850T models only

1   To gain access to the rear brake assembly remove the rear wheel as described in Chapter 5, Section 8.2-8.6. Lift the brake backplate out of the hub as a complete unit.
2   Inspect the brake linings for wear. Thin linings should be renewed after removing the two shoes from the brake backplate. Inspect the surface of the drum for scoring and ovality. The former fault will cause accelerated lining wear and reduced braking efficiency. The latter fault will cause brake judder.
3   After removal of the brake shoes the operating cam should be detached. Loosen the pinch bolt on the cam shaft and pull off the arm. Push the cam shaft from place. Clean and relubricate the cam shaft before refitting. Apply a small amount of heavy grease to the cam lobes.
4   If the brake shoe return springs are deformed or the spring ends are worn, the springs should be renewed as a pair.

### 14 Rear wheel bearings: examination and adjustment

1   On all but 750S and 850T models removal and examination of the rear wheels is similar to that given in Section 10 of this Chapter for the front wheel bearings. As with the front wheel, the left-hand bearing is retained by a circlip.
2   Models fitted with drum rear brakes utilise two tapered roller wheel bearings which allow adjustment to be made to take up normal wear. If discernable play can be felt at the wheel rim, the bearings should be removed for inspection, adjustment and regreasing.
3   Remove the wheel and lift the brake backplate assembly from the hub. Pull out the wheel spacers from the oil seals. The oil seals must be prised from position, using a screwdriver blade. If great care is taken not to distort the housings or damage the sealing lips, the seal can be reused. Lift out the bearing inner races. Do not allow the races to become interchanged; if they are to be refitted, they must be replaced in the same positions. Remove the bearing spacer.

4   Wash all the components, including the outer races in the hub, in petrol and then dry them thoroughly. Check the roller tracks and the rollers for pitting or scoring. If wear is evident the outer races may be drifted out using a suitable long punch inserted from the opposite side of the hub. If slight bearing wear was evident with the wheel on the machine, but no shims were found between the left-hand bearing inner race and the bearing spacers, the bearing must be renewed. New bearings or old bearings with shims already fitted should be adjusted, using shims available for this purpose.
5   Reassemble all the components, except the oil seals in the hub, with a shim between the left-hand bearing and bearing spacer to give zero end play in the bearings. Use the wheel spindle and a suitable tubular spacer to tighten the assembly together. Some experimentation will be required to arrive at the correct shim size. Dismantle the bearing assembly and add one 0.10 mm (0.019 in) shim to those already selected. When the bearings are refitted finally, the correct endfloat will automatically be provided.
6   Repack the bearings with a high melting point grease and replace all the components by reversing the dismantling procedure. The wheel can now be refitted to the machine.

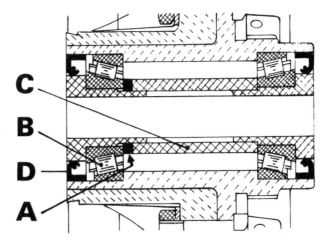

Fig. 6.6. Rear wheel bearing adjustment - 750S and 850T

A  =  Shims as required          C  =  Bearing spacer
B  =  Bearing - 2 off            D  =  Oil seal - 2 off

## 15 Cush drive assembly: removal, examination and replacement

1   A cush drive assembly is fitted to all but the 750S and 850T
models, to absorb snatch loads in the transmission. The assembly consists of a floating plate with a splined boss which locates
with the output boss on the rear bevel drive box. Webs on the
inner face of the plate locate with rubber buffers placed in the
rear wheel hub, which are prevented from rotation by webs cast
into the hub. After considerable service the rubber inserts will
compact, giving a noticeable backlash before the drive is taken
up.
2   Access to the cush drive unit can be made after removal of
the rear wheel from the frame. Remove the locking plate and

prise the large circlip from the wheel hub. A large screwdriver
is ideal for this operation. It should be possible to lift  the
cush drive plate straight off the steel sleeve in the centre
of the hub. No provision is made for lubricating the hub sleeve
and cush drive boss working surfaces after assembly. Because
of this, the sleeve and boss may have rusted, making dismantling
difficult. Apply a penetrating fluid to the small gap between the
sleeve and boss, and allow the fluid time to penetrate.
3   After removal of the drive plate lift out the twelve rubber
buffer inserts. These should be discarded if they are compacted
or are disintegrating. Reassemble the cush drive by reversing the
dismantling procedure. Ensure that the sleeve and boss are well
greased. The stepped side of the circlip must face upwards.

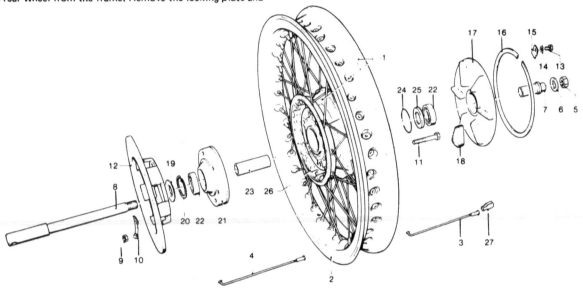

Fig. 6.7. Rear wheel and cush drive - Except 750S and 850T

| | | | |
|---|---|---|---|
| 1  Rear wheel | 8   Wheel spindle | 15  Lock plate | 22  Journal ball bearing - 2 off |
| 2  Rim | 9   Nut - 6 off | 16  Circlip | 23  Journal ball bearing - 2 off |
| 3  RH spoke and nipple - 20 off | 10  Lock plate - 3 off | 17  Cush drive plate | 24  'O' ring |
| 4  LH spoke and nipple - 20 off | 11  Bolt - 6 off | 18  Rubber insert - 12 off | 25  Cup |
| 5  Nut | 12  Brake disc | 19  Spacer | 26  Hub |
| 6  Washer | 13  Bolt | 20  Circlip | 27  Balance weight |
| 7  Spacer | 14  Star washer | 21  Disc spacer | |

15.2a Remove the locking bolt and plate and circlip

15.2b Lift off the cush drive plate

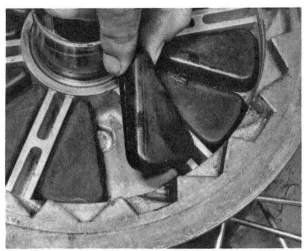

15.3a Cush drive consists of twelve separate inserts

15.3b Grease the cush drive boss heavily before fitting the plate

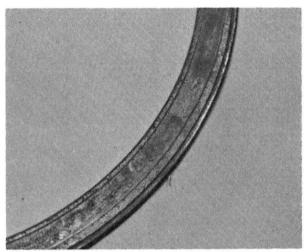

15.3c Shouldered face of circlip must face outwards

## 16 Parking brake: adjustment and maintenance

1    A mechanical caliper is fitted to the rear brake caliper bracket, to the rear of the standard hydraulic caliper, to act as a parking brake when the machine is at rest. The caliper is operated by the prop stand via a cable.

2    If the brake does not prevent movement of the machine when the stand is extended, adjustment may be made by means of the adjuster screw at the forward end of the cable.

3    Because the brake pads are rarely subjected to wear caused by the rotating disc, they will have a very long life, probably that of the machine itself. Maintenance is minimal, requiring only lubrication of the pivot points on the mechanical linkage at the caliper and of the pivot and push bar arrangement at the propstand.

16.1a Adjustment of parking brake cable - V1000 models

16.1b Mechanical caliper retained by two bolts

## 17 Tyres: removal and replacement

1  At some time or other the need will arise to remove and replace the tyres, either as the result of a puncture or because a renewal is required to offset wear. To the inexperienced, tyre changing represents a formidable task yet if a few simple rules are observed and the technique learned, the whole operation is surprisingly simple.

2  To remove the tyre from the wheel, first detach the wheel from the machine by following the procedure in Chapter 5, Section 2.7 for the front wheel and Section 8, paragraphs 5 and 6 for the rear wheel. Deflate the tyre by removing the valve insert and when it is fully deflated, push the bead of the tyre away from the wheel rim on both sides so that the bead enters the centre well of the rim. Remove the locking cap and push the tyre valve into the tyre itself.

3  Insert a tyre lever close to the valve and lever the edge of the tyre over the outside of the wheel rim. Very little force should be necessary; if resistance is encountered it is probably due to the fact that the tyre beads have not entered the well of the wheel rim all the way round the tyre.

4  Once the tyre has been edged over the wheel rim, it is easy to work around the wheel rim so that the tyre is completely free on one side. At this stage, the inner tube can be removed.

5  Working from the other side of the wheel ease the other edge of the tyre over the outside of the wheel rim furthest away. Continue to work around the rim until the tyre is free from the rim.

6  If a puncture has necessitated the removal of the tyre, reinflate the inner tube and immerse it in a bowl of water to trace the source of the leak. Mark its position and deflate the tube. Dry the tube and clean the area around the puncture with a petrol-soaked rag. When the surface has dried, apply rubber solution and allow this to dry before removing the backing from the patch and applying the patch to the surface.

7  It is best to use a patch of the self-vulcanising type, which will form a very permanent repair. Note that it may be necessary to remove a protective covering from the top surface of the patch, after it has sealed in position. Inner tubes made from synthetic rubber may require a special type of patch and adhesive, if a satisfactory bond is to be achieved.

8  Before replacing the tyre, check the inside to make sure the agent that caused the puncture is not trapped. Check also the outside of the tyre, particularly the tread area, to make sure nothing is trapped that may cause a further puncture.

9  If the inner tube has been patched on a number of past occasions, or if there is a tear or large hole, it is preferable to discard it and fit a new one. Sudden deflation may cause an accident, particularly if it occurs with the front wheel.

10  To replace the tyre, inflate the inner tube sufficiently for it to assume a circular shape but only just. Then push it into the tyre so that it is enclosed completely. Lay the tyre on the wheel at an angle and insert the valve through the rim tape and the hole in the wheel rim. Attach the locking cap on the first few threads, sufficient to hold the valve captive in its correct location.

11  Starting at the point furthest from the valve, push the tyre bead over the edge of the wheel rim until it is located in the central well. Continue to work around the tyre in this fashion until the whole of one side of the tyre is on the rim. It may be necessary to use a tyre lever during the final stages.

12  Make sure there is no pull on the tyre valve and again commencing with the area furthest from the valve, ease the other bead of the tyre over the edge of the rim. Finish with the area close to the valve, pushing the valve up into the tyre until the locking cap touches the rim. This will ensure the inner tube is not trapped when the last section of the bead is edged over the rim with a tyre lever.

13  Check that the inner tube is not trapped at any point. Reinflate the inner tube, and check that the tyre is seating correctly around the wheel rim. There should be a thin rib moulded around the wall of the tyre on both sides, which should be equidistant from the wheel rim at all points. If the tyre is unevenly located on the rim, try bouncing the wheel when the tyre is at the recommended pressure. It is probable that one of the beads has not pulled clear of the centre well.

14  Always run the tyres at the recommended pressures and never under or over-inflate. The correct pressures for solo use are given in the Specifications Section of this Chapter.

15  Tyre replacement is aided by dusting the side walls, particularly in the vicinity of the beads, with a liberal coating of French Chalk. Washing-up liquid can also be used to good effect, but this has the disadvantage of causing the inner surfaces of the wheel rim to corrode.

16  Never replace the inner tube and tyre without the rim tape in position. If this precaution is overlooked there is good chance of the ends of the spoke nipples chafing the inner tube and causing a crop of punctures.

17  Never fit a tyre that has a damaged tread or side walls. Apart from the legal aspects, there is a very great risk of blow-out, which can have serious consequences on any two-wheeled vehicle.

18  Tyre valves rarely give trouble, but it is always advisable to check whether the valve itself is leaking before removing the tyre. Do not forget to fit the dust cap, which forms an effective second seal.

## 18 Valve cores and caps

1  Valve cores seldom give trouble, but do not last indefinitely. Dirt under the seating will cause a puzzling 'slow-puncture'. Check that they are not leaking by applying spittle to the end of the valve and watching for air bubbles.

2  A valve cap is a safety device, and should always be fitted. Apart from keeping dirt out of the valve, it provides a second seal in case of valve failure, and may prevent an accident resulting from sudden deflation.

## 19 Front wheel balancing

1  The front wheel should be statically balanced, complete with tyre. An out of balance wheel can produce dangerous wobbling at high speed.

2  Some tyres have a balance mark on the sidewall. This must be positioned adjacent to the valve. Even so, the wheel still requires balancing.

3  With the front wheel clear of the ground, spin the wheel several times. Each time, it will probably come to rest in the same position. Balance weights should be attached diametrically opposite the heavy spot, until the wheel will not come to rest in any set position, when spun.

4  Balance weights, which clip round the spokes, are available in 5, 10 or 20 gramme weight. If they are not available, wire solder wrapped round the spokes and secured with insulating tape will make a substitute.

5  It is possible to have a wheel dynamically balanced at some dealers. This requires its removal.

6  There is no need to balance the rear wheel under normal road conditions, although any tyre balance mark should be aligned with the valve.

7  Machines fitted with cast aluminium wheels require special balancing weights which are designed to clip onto the centre rim flange, much in the way that weights are affixed to car wheels.

**20 Fault diagnosis: wheels, brakes and tyres**

| Symptom | Cause | Remedy |
|---|---|---|
| Handlebars oscillate at low speeds | Buckled front wheel | Remove wheel for specialist attention. Renew wheel (cast alloy type). |
| | Incorrectly fitted front tyre | Check whether line around bead is equidistant from rim. |
| Forks 'hammer' at high speeds | Front wheel out of balance | Add weights until wheel will stop in any position. |
| Brakes feel spongy | Air in hydraulic line<br>Fluid leak in system | Bleed brakes.<br>Renew faulty part. |
| Tyres wear more rapidly in middle of tread | Over-inflation | Check pressures and run at recommended settings. |
| Tyres wear rapidly at outer edges of tread | Under-inflation | Check pressures and run at recommended settings. |

# Chapter 7 Electrical system

## Contents

## Specifications

### Battery

| | 750S and 850T | 750S3 and 850T3 V-1000 and Le Mans |
|---|---|---|
| Type ... ... ... ... ... ... ... ... ... | M | 6DS 11 |
| Make ... ... ... ... ... ... ... ... | Marelli | Marelli |
| Capacity ... ... ... ... ... ... ... ... | 32 ah | 32 ah |
| Polarity ... ... ... ... ... ... ... ... | Negative earth | Negative earth |

### Alternator

| | | |
|---|---|---|
| Make ... ... ... ... ... ... ... ... | Bosch | Bosch |
| Type ... ... ... ... ... ... ... ... | G1 (R) 14V | G1 14V 20A 21 |
| Output ... ... ... ... ... ... ... ... | 14V/180W | 14V/280W |
| Charging starts at ... ... ... ... ... ... | 980 rpm | 1000 rpm |
| Field winding resistance ... ... ... ... ... | 6.90 ohms + 10% | |

### Voltage regulator

| | |
|---|---|
| Make ... ... ... ... ... ... ... ... ... | Bosch |
| Type ... ... ... ... ... ... ... ... | AD/1/14V |
| Regulated voltage: | |
| Under load ... ... ... ... ... ... ... | 13.9 - 14.8V |

### Rectifier

| | |
|---|---|
| Make ... ... ... ... ... ... ... ... | Bosch 14V/15A |

### Starter motor

| | | |
|---|---|---|
| Make ... ... ... ... ... ... ... ... | Bosch | Bosch |
| Type ... ... ... ... ... ... ... ... | DG(L) (DF, 850T model) | DF |
| Output ... ... ... ... ... ... ... ... | 0.4 HP (0.5 HP, 850T model) | 0.6 HP |
| Commutator minimum diameter ... ... ... ... | 31.2 mm (1.2283 in) | 33 mm (1.299 in) |
| Minimum brush length ... ... ... ... ... | 11.5 mm (0.4527 in) | |
| Armature axial float ... ... ... ... ... | 0.05 - 0.2 mm (0.002 - 0.008 in) | |

### Horns

| | | |
|---|---|---|
| Make ... ... ... ... ... ... ... ... | Belli | Belli |
| High tone current ... ... ... ... ... ... | 3A | 3A |
| Low tone current ... ... ... ... ... ... | 4A | 4A |

## Bulbs

| | | | | | | | | | | |
|---|---|---|---|---|---|---|---|---|---|---|
| Headlamp, European | ... | ... | ... | ... | ... | ... | ... | 40/45W | 40/45W |
| Headlamp, USA | ... | ... | ... | ... | ... | ... | ... | 40/45W sealed beam | 40/45W |
| Parking | ... | ... | ... | ... | ... | ... | ... | ... | 5W (European only) | 3W |
| Tail/stop | ... | ... | ... | ... | ... | ... | ... | ... | 5/21W | 5/21W |
| Oil pressure | ... | ... | ... | ... | ... | ... | ... | 1.2W | 1.2W |
| Neutral indicator | ... | ... | ... | ... | ... | ... | ... | 1.2W | 1.2W |
| Alternator charge | ... | ... | ... | ... | ... | ... | ... | 1.2W | 1.2W |
| Main beam | ... | ... | ... | ... | ... | ... | ... | — | 1.2W |
| Parking light warning | ... | ... | ... | ... | ... | ... | ... | — | 1.2W |
| Instrument lighting | ... | ... | ... | ... | ... | ... | ... | 3W | 3W |
| Flashing indicator | ... | ... | ... | ... | ... | ... | ... | 21W | 21W |
| Under seat illumination lamp | ... | ... | ... | ... | ... | ... | 3W (750S model only) | |

| | | | | | | | | | |
|---|---|---|---|---|---|---|---|---|---|
| **Fuses** | ... | ... | ... | ... | ... | ... | ... | 8 at 15 amps) | 6 at 16 amps |
| | | | | | | | | 1 at 25 amps) 750S | |
| | | | | | | | | 6 at 16 amps    850T | |

## 1 General description

The charging system consists of a 12 volt three-phase alternator driven directly by the crankshaft. The alternator has a rotor fixed to the front of the crankshaft, rotating within a stator screwed to the crankcase. The rotor is not permanently magnetised, energising current being supplied by the charge warning lamp circuit, via two brushes bearing on slip rings.

The a.c. voltage generated is rectified to d.c. by the diodes on the plate attached to the frame below the right-hand frame side cover, and regulated by the mechanical voltage regulator mounted below the petrol tank on the top frame tubes.

## 2 Battery: removal

1   Lift the dualseat and secure it in the upright position. Remove the tool tray and detach the battery retaining straps. Disconnect the battery leads by unscrewing the terminal bolts; always remove the negative lead (Earth) first.
2   The battery can now be lifted out. In addition to being very heavy - the battery is unusually large - the battery is retained in a confined space. Lift the battery up at the rear and slide it from position at about 45°. Spillage of the electrolyte is unlikely but if it does occur, wash the acidic fluid off with plenty of cold water, to prevent corrosion of the affected parts.

2.1 Battery is vast and very heavy

## 3 Battery: maintenance

1   The battery fitted to all the Moto Guzzi V-twins covered in this manual is rated at 32 ah. Inspection and replenishment of the fluid level may be made through the single filler orifice in the battery top, after prising out the filler plug. Inspection may be aided by the use of a small torch. **DO NOT** hold a naked flame near the cell filler hole; the electrolyte gives off hydrogen and oxygen (a notably inflammable mixture).
2   Top up the battery with distilled water if the level is below that of the plates. Do not fill more than 5 mm (3/16 in) above the level of the plates. If a battery is used having separate cell filler caps, check each cell individually and replenish as required. The condition of the battery may be checked using a small hydrometer. The reading should be from 1.260 - 1.280. If the specific gravity is lower than this, the battery should be charged from a trickle charger.
3   The normal safe charging rate for a battery is 1/20th of the battery capacity. The charge rate for a 32 amp/hour battery is therefore approximately 1½ - 2 amps. Charging the battery at a slightly higher level is permissible (up to 3 amps) but this may shorten the life of the battery and should therefore be avoided if at all possible.
4   Ensure that the battery lead connections are clean and tight. Apply petroleum jelly to the terminals to prevent corrosion. Ensure that the battery is earthed well.
5   All models have a negative earth system.
6   If the motorcycle is not going to be used for a time, the battery should be put on charge every six weeks. If the battery is permitted to discharge completely the plates will sulphate, indicated by a grey colour, and render the battery useless. A fully charged battery will have plates a muddy brown colour. If the case has a sediment on the bottom, the plates are breaking up and the battery will soon require replacement. Disconnect the battery from the motorcycle electrical system if it is being charged on the machine, or the rectifier diodes will suffer. Remove the cell filler plugs when charging.

## 4 Crankshaft alternator: checking the output

1   The output from the alternator mounted on the end of the crankshaft can be checked only with specialised test equipment of the multi-meter type. It is unlikely that the average owner/rider will have access to this equipment or instruction in its use in consequence, if the performance of the alternator is in any way suspect, it should be checked by a Moto Guzzi agent or an auto-electrical specialist.

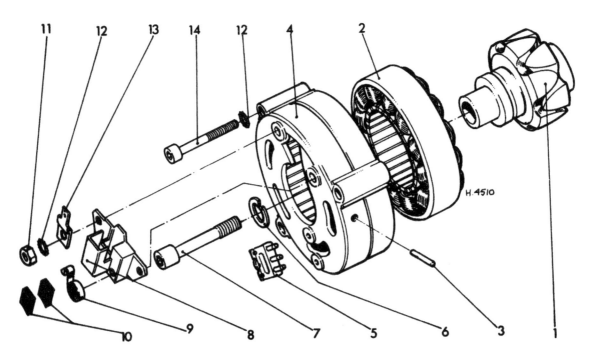

**Fig. 7.1. Alternator**

| | | | | | |
|---|---|---|---|---|---|
| 1 | Rotor - 1 off | 5 | Three pin connector - | 8 | Brush holder - 1 off |
| 2 | Stator - 1 off | | 1 off | 9 | Brush spring - 2 off |
| 3 | Spring dowel pin - 2 off | 6 | Spring washer - 1 off | 10 | Brush - 2 off |
| 4 | Alternator housing - 1 off | 7 | Rotor centre screw - 1off | 11 | Nut - 2 off |

| | |
|---|---|
| 12 | Spring washer - 9 off |
| 13 | Terminal - 2 off |
| 14 | Socket cap screw,- 3 off |

## 5  Alternator: removal

1   Disconnect the battery negative (earth) lead.
2   Slacken the horn fixing screws.
3   Unscrew the three hexagon socket cap screws and remove the engine front cover.
4   Pull off the three-pin output socket from the alternator plug. Pull off the two leads to the brush holder. Note the colours of the leads for replacements.
5   Pull up the alternator brushes and wedge them clear of the slip rings with the ends of the brush springs.
6   Unscrew the three hexagon socket cap screws, and remove the alternator stator.
7   Unscrew the rotor centre fastening screw with an Allen key. An extractor is required to release the rotor from its taper. To provide a bearing for the extractor centre screw, replace the rotor fastening screw. The rotor is threaded and when the fastening screw reaches the end of the thread it will be felt to be loose, until it engages with the thread on the end of the crankshaft. Engage the first few threads. Alternatively, make a headed pin that will pass through the rotor, and contact the end of the crankshaft. It should be 3/16 in diameter and 2¼ in long. Tighten the extractor centre screw and tap gently if necessary to break the taper. Remove the rotor.

## 6  Alternator: renovation

1   Check that the brushes move freely in their holders and that the springs press them firmly against the slip rings. The brushes should be renewed if they are worn badly.
2   The two nuts on the brush holder, visible from the front of the alternator stator, retain the blade terminals only.
3   To replace the brushes, unscrew the two nuts accessible from

inside the stator housing, remove with spring washers and take off the brush housing. Note the insulating washers and bush on the right-hand stud (viewed from the front of the housing).
4   When soldering new brushes in position, do not allow solder to run down the brush tails towards the brushes. Note that the brush tails locate in slots behind the brushes.
5   Clean dirty slip rings with petrol, or if necessary very fine glass paper. **Do not** use emery paper. Scored slip rings must be skimmed in a lathe to a minimum diameter of 26.8 mm (1.055 in).
6   The stator and rotor winding have to be checked for short circuits with 40 volts a.c., this requires specialised equipment. The resistance between phase outputs may also be checked. Set the multi-meter to read ohms. Check the resistance between each pair of alternator output terminals on the three-pin plug. The value should be 0.62 ohms. Check the resistance of the energiser winding across the slip rings, the value should be 6.90 ohms + 10%.

## 7  Diode plate rectifier: removal

1   The diode plate is mounted behind the right-hand side cover. On the plate are fitted the diodes which rectify the three phase a.c. output from the alternator, to d.c. for charging the battery.
2   Checking the diodes requires specialised equipment not normally available to the amateur mechanic. If one diode is faulty, the complete assembly must be replaced.
3   Disconnect the battery negative lead. Remove the right-hand side cover and disconnect the multiple socket and two separate leads to the rectifier plate. Unscrew the four retaining screws and lift the plate away sufficiently to disconnect the final leads.
4   It is imperative that the leads are connected correctly on refitting the plate. To this end note the wire positions carefully during removal.

5.4a Note alternator wiring before removal

5.4b Wedge the brushes by means of the springs and ...

5.6 ... lift the stator from position

5.7a Remove the central bolt and ...

5.7b ... lift the rotor off the tapered shaft

6.1 Check the condition and length of brushes

### 8  Voltage regulator: removal

1  The voltage regulator, mounted below the top frame tubes, is not adjustable. If it is suspected of being faulty, it should be checked by a specialist.
2  To remove, unplug the three-pin plug, unscrew the two screws and lift the unit away.

### 9  Headlamp: replacing the bulbs and adjusting the beam

1  Remove the single retaining screw from the base of the headlamp rim. On Le Mans models removal of the small handlebar fairing is necessary to gain access to the screw. On 750S models the rim is split at the lower edge and is secured by a clamping bolt. Lift the rim up and out from the lower edge and away from the shell.
2  The headlamp bulb is retained in a holder which is secured to the rear of the reflector unit by two coil springs. Unhook the springs and pull the holder from position. If required, pull off the lead socket from the pins at the rear of the holder. The bulb has an offset pin bayonet fixing. Push the bulb in, twist it to the left and release, to remove it. The pins are offset, to prevent replacement of the bulb in the wrong position.
3  The pilot bulb (where fitted) is also bayonet fixed, in a

holder which is a push fit in the reflector unit. Before replacing bulbs, ensure that the contacts are clean.

When replacing the headlamp, hook the rim over the top of the shell and push the bottom of the rim over the retaining clips at the bottom of the shell. Ensure that it is firmly fitted, since it does not bounce!
4  On 750S models, headlamp beam height is adjusted by slackening the two headlamp fixing bolts fixing the lamp to the fork brackets and pivoting the headlamp in a vertical plane. All other models are fitted with three screws passing through the headlamp rim at equidistant intervals around the periphery. By loosening or tightening the screws the headlamp beam may be adjusted within limits in any plane. Further vertical adjustment may be made by means of the headlamp shell pivot bolts. When adjusting the beam tyre pressures should first be checked and the rider should be seated normally.

UK lighting regulations stipulate that the lighting system must be arranged so that the light does not dazzle a person standing in the same horizontal plane as the vehicle, at a distance greater than 25 feet from the lamp, whose eye level is not less than 3 feet 6 inches above that plane. It is easy to approximate this setting by placing the machine 25 feet away from a wall, on a level road and setting the beam height so that it is concentrated at the same height as the distance from the centre of the headlamp to the ground. In addition, the headlamp must be capable of being dipped.

7.1 Rectifier mounted behind right-hand frame cover

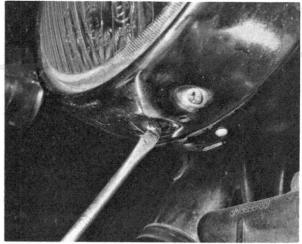

9.1 Headlamp rim held by single screw - except 750 models

9.2a Detach springs at rear of holder to ...

9.2b ... allow removal of headlamp bulb

9.3 Pilot bulb is a pushfit in reflector

## 10 Indicator relay: replacing

1    The indicator relay is mounted in a rubber holder to isolate·
the unit from the effects of vibration. The relay is located forward
of the rectifier.
2    The flashing rate should be between 60 and 120 times per
minute in the UK. If the rate deviates, check the indicator bulb
wattages and contacts, or wiring - particularly earth returns, or
switch.
3    If the relay malfunctions, the usual indication is one flash
before the system goes dead. The relay will require renewal if the
fault cannot be traced to a bulb or wiring, or the indicator
switch. Handle the relay with care, as it is easily damaged by
being dropped.

## 11 Fuses: replacing

1    The six fuses are contained in a bank within a plastic box
below the right-hand side cover. The box lid is secured by a
knurled bolt, Each of the fuses are interconnected permanently,
none being fitted as a spare.
     A blown fuse can be recognised by the melted metal strip. If
a fuse blows repeatedly, the electrical system should be checked
to eliminate the fault. Do not put in a fuse of a higher rating -
another item may be damaged - or a fire result. The same
applies to replacing the fuse with wire. Spare fuses of the correct
rating should always be carried.

## 12 Instrument lighting and warning lamps: replacing bulbs

1    Both speedometer and rev-counter are internally illuminated.
In addition there are a number of warning lights fitted to the
lighting console integral with the instrument housing.
2    All bulbs are of the bayonet type fitted into holders which
are a push fit in the underside of the instruments or lighting
panel.

## 13 Indicators and rear/stop lamp: replacing bulbs

1    The indicators and rear lamp are similar in design. Both lens
and reflector with integral bulb holder are retained in the
housing by two screws. Unscrew the screws, remove the lens and
reflector and separate. Check the condition of the gasket.

2    Check the electrical connections on the rear of the reflector,
particularly the earths.
3    The bulbs are bayonet type, the rear/stop lamp bulb has
double filaments, with off-set pins to ensure correct orientation.
4    The V-1000 model has in addition, a separate number plate
illuminating unit. The lens cover is retained by two screws.
5    Always replace indicator bulbs with those of the correct
rating, or the flashing rate will be upset.

## 14 Ignition switch

1    The ignition switch is mounted forward of the fuel tank, on
the top frame tube. If the condition of the switch is suspect, a
battery and bulb or a multi-meter may be used to check
continuity with the unit removed from the machine.
2    After removal of the petrol tank in order to gain access,
remove the two bolts holding the switch to the frame. Disconnect the wiring leads by pulling out the socket. The switch unit
proper may be withdrawn from the mounting sleeve, by pressing
in the spring loaded peg with a pointed instrument. Repair of a
faulty switch is impracticable; a new unit together with the
correct key should be acquired.

## 15 Stoplamp switches and brake fluid level switch

1    There are front and rear brake operated stoplamp switches,
neither can be repaired.
2    If a faulty switch is suspected, check for continuity.
3    On drum rear brake models the switch is a mechanical unit
actuated by the brake pedal. On all hydraulic brakes the switches
are incorporated in the hydraulic hose junction boxes. A faulty
switch should be renewed.
4    The brake fluid level indicator switch, fitted to V-1000 and
Le Mans models, is located in the fluid reservoir cap. Again, it
cannot be repaired, the cap must be replaced. The warning lamp
may flash under heavy braking, but this does not indicate a
fault.

## 16 Prop stand switch: V-1000 Convert models only

1    On V-1000 models, a switch is fitted, operated by the propstand, which prevents functioning of the ignition system when
the stand is extended. In addition, the switch operates a warning
light which indicates that the stand has not been retracted.
2    The switch is mounted on a bracket forward of the prop
stand bracket. Continuity of the switch may be checked by using
a battery and bulb.

11.1 Fuses retained in bank within plastic box

12.2 Instrument illuminating bulb holders are push fit

13.1 Separate rear light and number plate light - V-1000 models

13.2 Rear light lens held by two screws as are ...

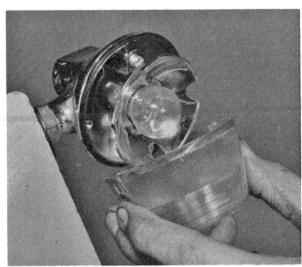

13.3 ... lens covers on flashing indicators

15.3 Stop light switches included in hose junction units

15.4a Fluid level switch on V-1000 models is ...

15.4b ... operated by means of a float

17.1 Starter isolator is included in clutch cable

## 17 Starter isolating switch

1   A switch is incorporated in the clutch cable on all but Le Mans models to prevent engine starting when the clutch is not disengaged. The switch is a sealed unit and is integral with the clutch cable. Consequently if failure occurs the clutch cable must be renewed as a unit.

2   If the switch malfunctions and does not complete the starter circuit the safety feature may be overriden temporarily by disconnecting the two leads to the switch and joining them together.

## 18 Horns

1   Two horns, of useful volume, are fitted to all models. The high tone horn has a current rating of 3 amperes, and the low tone unit a rating of 4 amperes.

2   On all models except V-1000 machines, the horns are mounted as a pair on the frame front down tubes. The horns on V-1000 models are mounted separately under the front crashbar air deflectors (stabilisers).

3   If both horns do not sound check the continuity of the switch and wiring. If one horn does not sound check the wiring and then check the horn from a separate power source. A faulty unit should be renewed: repair is not possible.

## 19 Wiring: layout and examination

1   The wiring is colour coded in accordance with the accompanying wiring diagrams.

2   Screw terminals and spring clip terminals should be checked for tightness. Inspect blade type terminals for good contact. Pay special attention to earth connections. If the lights are poor, or the bulbs blow frequently, the earth returns are probably poor, and connections should be inspected. Check the wires for damaged insulation, they must not be pulled tight, nor routed over sharp edges.

**20 Fault diagnosis: electrical system**

| Fault | Cause | Remedy |
|-------|-------|--------|
| Alternator making noises | Brushes squeaking | Renew brushes or clean slip rings. |
| Charge warning lamp glows at half strength when engine is idling | Poor contact in wiring<br>Faulty regulator<br>Worn brushes<br>Rectifier diode shorting<br>Rotor or stator shorting | Check wiring and connections.<br>Check and renew if necessary.<br>Renew.<br>Check and renew if necessary.<br>Check and renew, if necessary. |
| Battery overcharging | Poor contact between regulator and alternator<br>Faulty regulator | Check wiring.<br><br>Check and renew, if necessary. |
| Charge warning lamp remains alight or glows when engine speed is above idling | Faulty regulator<br>Poor contact in wiring<br>Worn brushes<br>Faulty rotor winding<br>Poor contact in rotor energising circuit<br>Faulty diodes | Check and renew if necessary.<br>Check wiring.<br>Renew.<br>Check and renew, if necessary.<br>Check wiring.<br>Check and renew, if necessary. |
| Charge warning lamp will not light when ignition is switched on | Faulty bulb.<br>Poor connection | Renew.<br>Check wiring. |
| Complete electrical failure | Blown fuse | Renew after tracing fault. |
| Dim lights, horn inoperative | Discharged battery | Check alternator output.<br>Check condition of battery. |
| Bulbs 'blow' | Vibration, poor contact | Check that bulb holders are secure.<br>Check earth connections. |

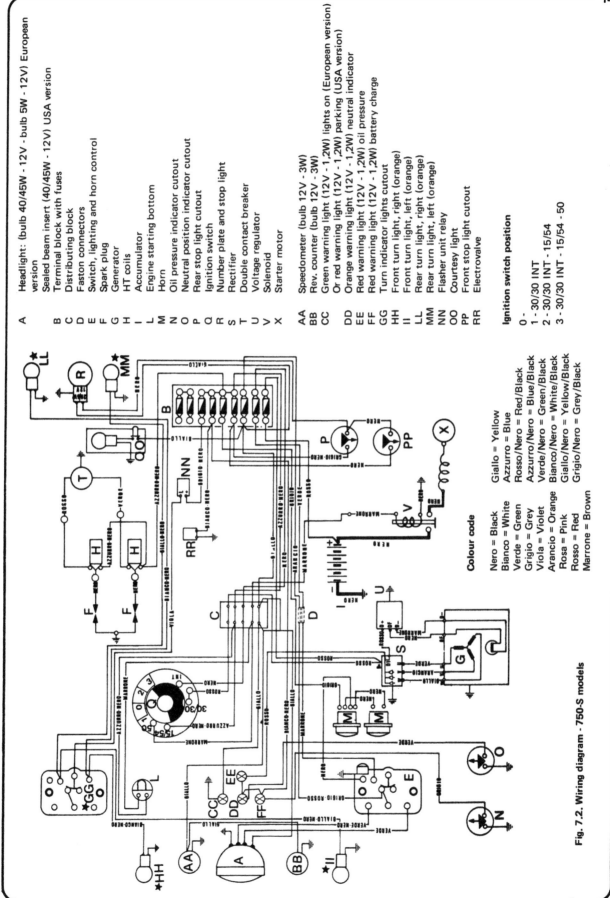

A — Headlight: (bulb 40/45W - 12V - bulb 5W - 12V) European version
B — Sealed beam insert (40/45W - 12V) USA version
C — Terminal block with fuses
D — Distributing block
E — Faston connectors
F — Switch, lighting and horn control
G — Spark plug
H — Generator
I — HT coils
L — Accumulator
M — Engine starting bottom
N — Horn
O — Oil pressure indicator cutout
P — Neutral position indicator cutout
Q — Rear stop light cutout
R — Ignition switch
S — Number plate and stop light
T — Rectifier
U — Double contact breaker
V — Voltage regulator
X — Solenoid
  — Starter motor

AA — Speedometer (bulb 12V - 3W)
BB — Rev. counter (bulb 12V - 3W)
CC — Green warning light (12V - 1,2W) lights on (European version)
DD — Or red warning light (12V - 1,2W) parking (USA version)
EE — Orange warning light (12V - 1,2W) neutral indicator
FF — Red warning light (12V - 1,2W) oil pressure
GG — Red warning light (12V - 1,2W) battery charge
HH — Turn indicator lights cutout
II — Front turn light, right (orange)
LL — Front turn light, left (orange)
MM — Rear turn light, right (orange)
NN — Rear turn light, left (orange)
OO — Flasher unit relay
PP — Courtesy light
RR — Front stop light cutout
  — Electrovalve

**Ignition switch position**

0 - 30/30 INT
1 - 30/30 INT
2 - 30/30 INT - 15/54
3 - 30/30 INT - 15/54 - 50

**Colour code**

Nero = Black
Bianco = White
Verde = Green
Grigio = Grey
Viola = Violet
Arancio = Orange
Rosa = Pink
Rosso = Red
Marrone = Brown

Giallo = Yellow
Azzurro = Blue
Rosso/Nero = Red/Black
Azzurro/Nero = Blue/Black
Verde/Nero = Green/Black
Bianco/Nero = White/Black
Giallo/Nero = Yellow/Black
Grigio/Nero = Grey/Black

**Fig. 7.2. Wiring diagram - 750-S models**

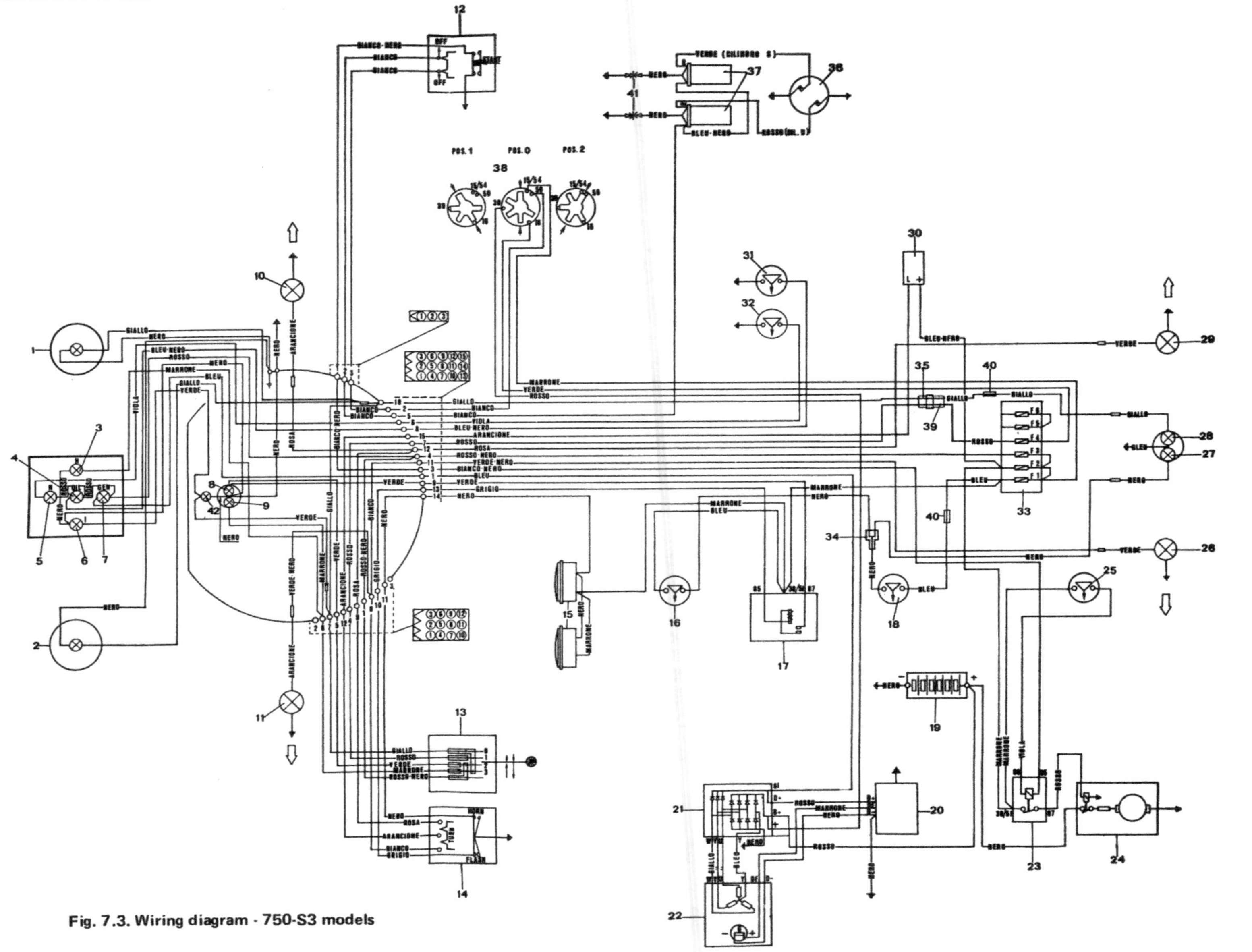

Fig. 7.3. Wiring diagram - 750-S3 models

**Fig. 7.3. Wiring diagram: 750 - S3 models**

1 Km counter
2 Rev. counter
3 High beam indicator light
4 Oil pressure indicator light
5 Neutral indicator light
6 Town driving indicator light
7 Generator charge indicator light
8 Low beam
9 High beam
10 Right front turn signal light
11 Left front turn signal light
12 Engine starting and stopping switch
13 Lighting switch
14 Switch: turn signal, starting, horns, flashing light
15 Horns (power 7 A)
16 Front brake stop light cutout
17 Flashing light relay
18 Rear brake stop light cutout
19 Battery (12V - 32 Ah)
20 Regulator
21 Rectifier
22 Alternator
23 Starter motor relay
24 Starter motor
25 Clutch cable cutout
26 Left rear turn signal
27 Rear brake stop light
28 Number plate and town driving light
29 Right rear turn signal
30 Flasher unit
31 Oil pressure cutout
32 Neutral position cutout
33 Terminal block with fuses (16 A)
34 3-way connector
35 4-way connector (AMP)
36 Contact breaker
37 Coils
38 Ignition switch (3 positions)
39 4-way connector (AMP)
40 2-way connector
41 Spark plugs
42 Town driving light, front

**Colour code**
Nero = Black
Bianco = White
Verde = Green
Grigio = Grey
Viola = Violet
Arancio = Orange
Rosa = Pink
Rosso = Red
Marrone = Brown

Giallo = Yellow
Azzurro = Blue
Rosso/Nero = Red/Black
Azzurro/Nero = Blue/Black
Verde/Nero = Green/Black
Bianco/Nero = White/Black
Giallo/Nero = Yellow/Black
Grigio/Nero = Grey/Black
Grigio/Rosso = Grey/Red

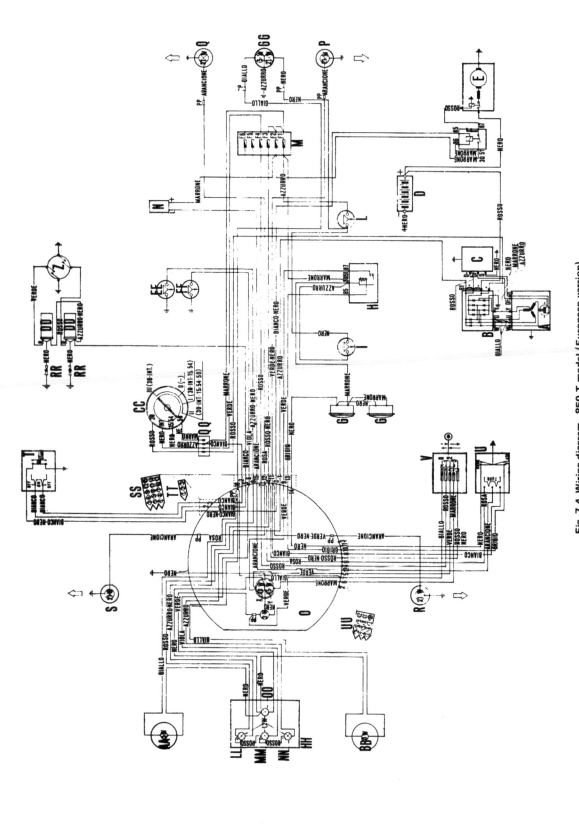

Fig. 7.4. Wiring diagram - 850-T model (European version)

**Colour code**

| Nero | = | Black |
|---|---|---|
| Bianco | = | White |
| Verde | = | Green |
| Grigio | = | Grey |
| Viola | = | Violet |
| Arancio | = | Orange |
| Rosa | = | Pink |
| Rosso | = | Red |
| Marrone | = | Brown |
| Giallo | = | Yellow |
| Azzurro | = | Blue |
| Rosso/Nero | = | Red/Black |
| Azzurro/Nero | = | Blue/Black |
| Verde/Nero | = | Green/Black |
| Bianco/Nero | = | White/Black |
| Giallo/Nero | = | Yellow/Black |
| Grigio/Nero | = | Grey/Black |
| Grigio/Rosso | = | Grey/Red |

| | |
|---|---|
| 1 | Km counter |
| 2 | Rev counter |
| 3 | High beam indicator light |
| 4 | Oil pressure indicator light |
| 5 | Neutral indicator light |
| 6 | Town driving indicator light |
| 7 | Generator charge indicator light |
| 8 | Low beam |
| 9 | High beam |
| 10 | Right front turn signal light |
| 11 | Left front turn signal light |
| 12 | Engine starting and stopping switch |
| 13 | Lighting switch |
| 14 | Switch; turn signal, starting, horns, flashing light |
| 15 | Horns power (7A) |
| 16 | Front brake stop light cutout |
| 17 | Flashing light relay |
| 18 | Rear brake stop light cutout |
| 19 | Battery (12V - 32 Ah) |
| 20 | Regulator |
| 21 | Rectifier |
| 22 | Alternator (14V - 20A) |
| 23 | Starter motor relay |
| 24 | Starter motor (12V - 0.7 HP) |
| 25 | Clutch cable cutout |
| 26 | Left rear turn signal |
| 27 | Rear brake stop light |
| 28 | Number plate and town driving light |
| 29 | Right rear turn signal |
| 30 | Flasher unit |
| 31 | Oil pressure cutout |
| 32 | Neutral position cutout |
| 33 | Terminal block with fuses (16A) |
| 34 | 3-way connector |
| 35 | 4-way connector (amp) |
| 36 | Contact breaker |
| 37 | Coils |
| 38 | Ignition switch (3 positions) |
| 39 | 4-way connector (amp) |
| 40 | 2-way connector |
| 41 | Spark plugs |
| 42 | Town driving light, front |

**Fig. 7.5. Wiring diagram: 850 - T3 model (European version)**

---

**Colour code**

| Nero | = Black |
|---|---|
| Bianco | = White |
| Verde | = Green |
| Grigio | = Grey |
| Viola | = Violet |
| Arancio | = Orange |
| Rosa | = Pink |
| Rosso | = Red |
| Marrone | = Brown |
| Giallo | = Yellow |
| Azzurro | = Blue |
| Rosso/Nero | = Red/Black |
| Azzurro/Nero | = Blue/Black |
| Verde/Nero | = Green/Black |
| Bianco/Nero | = White/Black |
| Giallo/Nero | = Yellow/Black |
| Grigio/Nero | = Grey/Black |

**Fig. 7.4. Wiring diagram: 850 - T model (European version)**

| | |
|---|---|
| A | Generator |
| B | Rectifier |
| C | Regulator |
| D | Battery |
| E | Starter motor |
| F | Starter motor relay |
| G | Horn |
| H | Flashing light relay |
| I | Hydrostop |
| L | Rear stop switch |
| M | Terminal block with fuses |
| N | Flasher unit |
| O | Asymmetric light |
| P | Left turn signal, rear |
| Q | Right turn signal, rear |
| R | Left turn signal, front |
| S | Right turn signal, front |
| T | Engine starter and stop switch |
| U | Control device, turn signals, horn, flashing light |
| V | Light switch: dimmer, city light, parking light |
| AA | Speedometer |
| BB | Rev. counter |
| CC | General commutator |
| DD | HT coil |
| EE | Oil light switch |
| FF | Neutral light switch |
| GG | Number plate and stop light |
| HH | Instrument panel |
| LL | Oil pressure light (red) |
| MM | Neutral light (orange) |
| NN | Battery light (red) |
| OO | City light (green) |
| QQ | 4-way connector (AMP) |
| RR | Spark plugs |
| SS | 15-way connector (MOLEX) |
| TT | 3-way connector (MOLEX) |
| UU | 12-way connector (MOLEX) |
| X | Low beam |
| Y | High beam |
| Z | Contact breaker |

**Fuses**

| F1 - 15A | Horn, stop, signals relay |
|---|---|
| F2 - 15A | Starter relay, flasher unit |
| F3 - 15A | Head light, lights LL; MM; NN |
| F4 - 15A | Parking light, light OO |
| F5 - 15A | Reserve |
| F6 - 15A | Reserve |

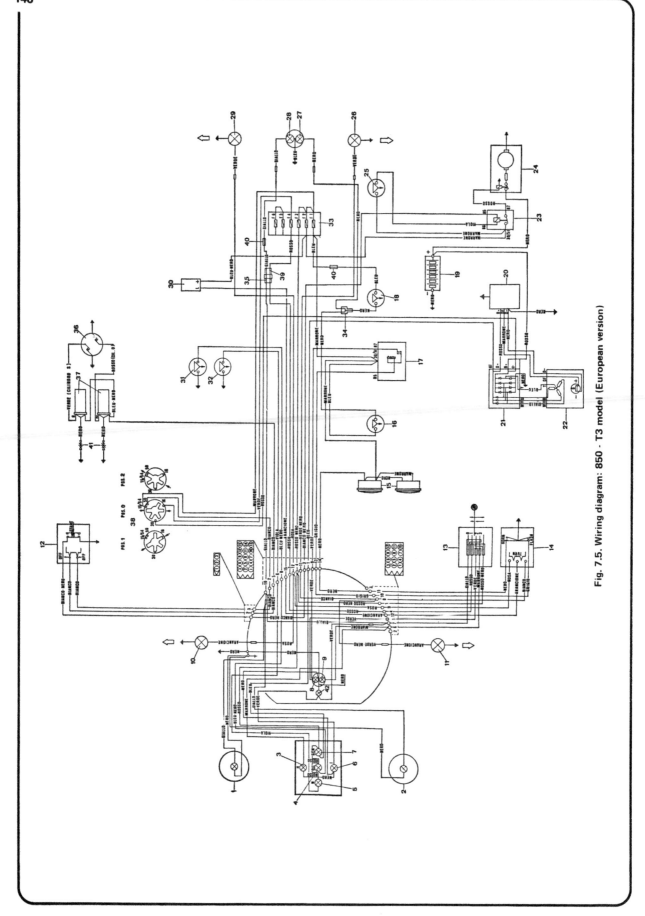

Fig. 7.5. Wiring diagram: 850 - T3 model (European version)

**Fig. 7.6. Wiring diagram: 850-T3 model (US version)**

1 Mile counter (bulb 3W)
2 Rev counter (bulb 3W)
3 High beam indicator light (1.2W)
4 Oil pressure indicator light (1.2W)
5 Neutral indicator (light 1.2W)
6 Low beam indicator light (1.2W)
7 Generator charge indicator light (1.2W)
8 Low beam (40W)
9 High beam (45W)
10 Right front turn signal light (21W)
11 Left front turn signal light (21W)
12 Engine starting and stopping switch
13 Lighting switch
14 Switch; turn signal, horns, flashing light
15 Horns power (7A)
16 Front brake stop light cutout
17 Flashing light relay
18 Rear brake stop light cutout
19 Battery
20 Regulator
21 Rectifier

22 Alternator
23 Starter motor relay
24 Starter motor
25 Clutch cable cutout
26 Left rear turn signal (21W)
27 Rear brake stop light (21W)
28 Number plate and parking light (5W)
29 Right rear turn signal (21W)
30 Flasher unit
31 Oil pressure cutout
32 Neutral position cutout
33 Terminal block with fuses (16A)
34 3-way connector
35 4-way connector
36 Contact breaker
37 Coils
38 Ignition switch (3 positions)
39 4-way connector
40 2-way connector
41 Spark plugs
42 Light switch, with stop device from position
   High-Low Beam to position Parking light

**Colour code**

| | | |
|---|---|---|
| Nero | = | Black |
| Bianco | = | White |
| Verde | = | Green |
| Grigio | = | Grey |
| Viola | = | Violet |
| Arancio | = | Orange |
| Rosa | = | Pink |
| Rosso | = | Red |
| Marrone | = | Brown |

| | | |
|---|---|---|
| Giallo | = | Yellow |
| Azzurro | = | Blue |
| Rosso/Nero | = | Red/Black |
| Azzurro/Nero | = | Blue/Black |
| Verde/Nero | = | Green/Black |
| Bianco/Nero | = | White/Black |
| Giallo/Nero | = | Yellow/Black |
| Grigio/Nero | = | Grey/Black |
| Grigio/Rosso | = | Grey/Red |

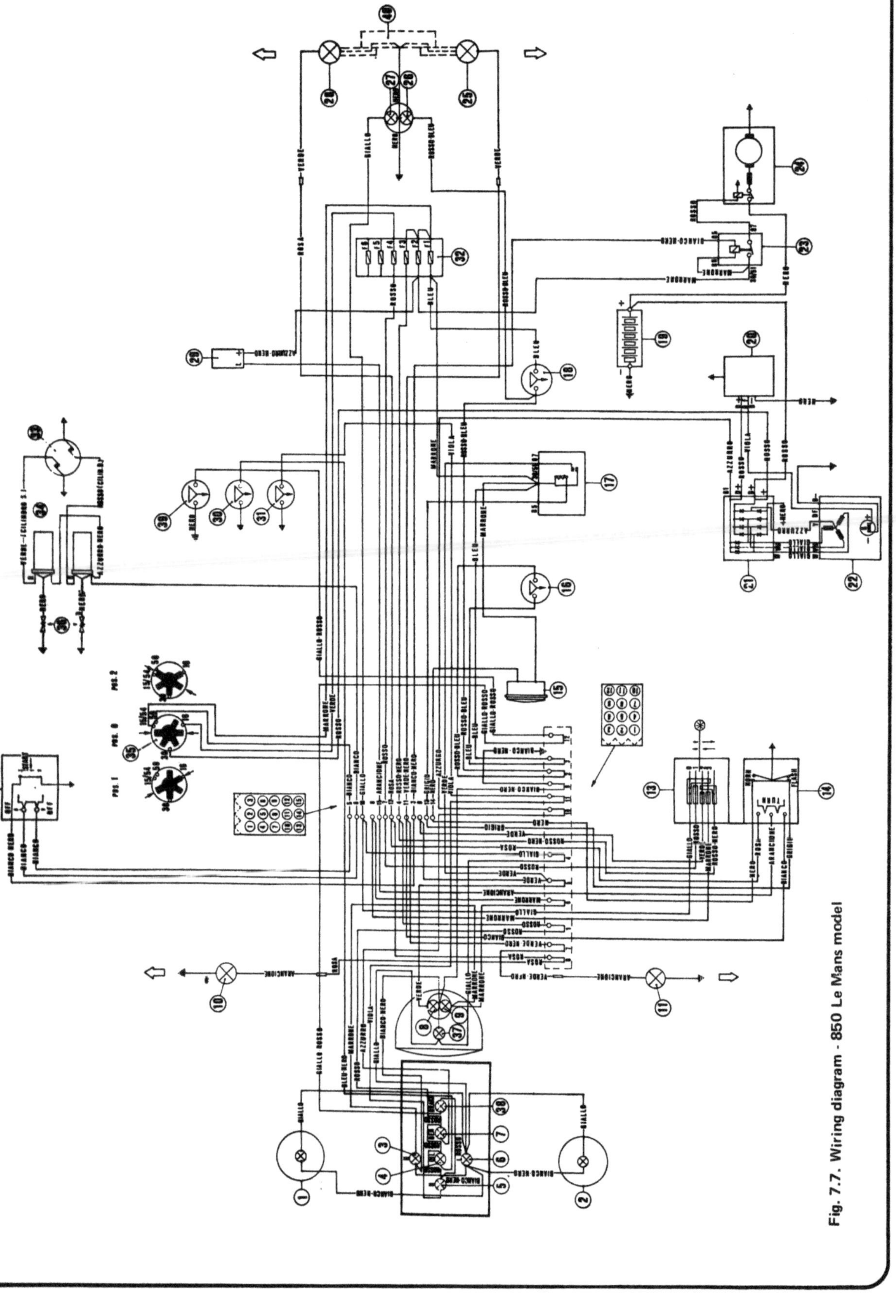

Fig. 7.7. Wiring diagram - 850 Le Mans model

**Fig. 7.7. Wiring diagram: 850 Le Mans model**

1 Km counter, bulb 3W
2 Rev counter, bulb 3W
3 High beam warning light, bulb 1.2W
4 Oil pressure warning light, bulb 1.2W
5 Neutral warning light, bulb 1.2W
6 Parking light warning light, bulb 1.2W
7 Generator charge warning light, bulb 1.2W
8 High beam    bulb 40/45W
9 Low beam
10 Right front turn signal, bulb 21W
11 left front turn signal, bulb 21W
12 Engine starting and stopping switch
13 Lighting switch
14 Switch turn signals, starting, horn, flashing light
15 Horn (Absorption 3.5A)
16 Front brake stop light cutout
17 Flashing light relay
18 Rear brake stop light cutout
19 Battery 12V - 20 Ah
20 Regulator
21 Rectifier
22 Alternator (14V 20A)
23 Starter motor relay
24 Starter motor (12V - 0.7 HP)
25 Left rear turn signal, bulb 21W
26 Rear brake stop light
27 Number plate and parking light    bulb 5/21W
28 Right rear turn signal, bulb 21W
29 Flasher unit
30 Oil pressure cutout
31 Neutral position cutout
32 Terminal block and fuses (16A fuses)
33 Contact breaker
34 Coils
35 Ignition switch (3 positions)
36 Spark plugs
37 Parking light front bulb 3W
38 Brake fluid level warning light (Brake) bulb 1.2W
39 Brake fluid level indicator cutout left and rear circuit

Colour code

Nero = Black
Bianco = White
Verde = Green
Grigio = Grey
Viola = Violet
Arancio = Orange
Rosa = Pink
Rosso = Red
Marrone = Brown
Giallo = Yellow
Azzurro = Blue
Rosso/Nero = Red/Black
Azzurro/Nero = Blue/Black
Verde/Nero = Green/Black
Bianco/Nero = White/Black
Giallo/Nero = Yellow/Black
Grigio/Nero = Grey/Black
Grigio/Rosso = Grey/Red

**Fig. 7.8. Wiring diagram: V1000 1 - Convert model (European version)**

1 Speedometer bulb (3W)
2 Additional light (bulb 5) - only on request
3 High beam warning light (bulb 1.2W) H
4 Oil pressure warning light (bulb 1.2W) Oil
5 Neutral position warning light bulb (1.2W) N
6 Town driving warning light (bulb 1.2W) L
7 Generator charge warning light (bulb 1.2W) Gen
8 Low beam    bulb 40/45W
9 High beam
10 Turn indicator light - front, right (bulb 21W)
11 Turn indicator light - left, front (bulb 21W)
12 Engine starting and stopping control
13 Additional light switch
14 Control: turn indicator lights, horns, flashing lights
15 Horns (consumption: 7A)
16 Front brake switch
17 Flashing light (flash) relay
18 Rear brake switch
19 Battery
20 Regulator
21 Rectifier
22 Alternator
23 Starter motor relay
24 Starter motor
25 Switch on clutch control wire
26 Turn indicator light - rear, left (bulb 21W)
27 Rear stop light (bulbs 5/21W)
28 Number plate light (bulb 5W)
29 Turn indicator light - rear, front (bulb 21W)
30 Turn indicator lights, flasher unit
31 Oil pressure switch (on the engine crankcase)
32 Town driving light, front (bulb 3W)
33 Terminal block with fuses (16A fuses)
34 3-way connector
35 4-way connector  Amp
36 Breaker
37 Coils
38 Ignition switch (3 positions)
39 Switch actuating rear turn indicator lights flashing
40 2-way connector
41 Spark plugs
42 Light switch with travel limit from position High/Low beam to position  Town driving light
43 Right turn indicator warning light (bulb 1.2W)
44 Left turn indicator warning light (bulb 1.2W)
45 Warning light indicating  Side stand in position Park (bulb 1.2W)
46 Brake fluid level warning light - brake (bulb 1.2W)
47 Fuel level warning light fuel (bulb 1.2W)
48 4-way connector (amp)
49 Connection
50 Brake fluid level indicator
51 Fuel level indicator
52 Electrovalve (2.5W)
53 Coil control device
54 Commutator for side stand warning light  Park position
55 Rear parking light (bulb 5/21W)

Colour code

Nero = Black
Bianco = White
Verde = Green
Grigio = Grey
Viola = Violet
Arancio = Orange
Rosa = Pink
Rosso = Red
Marrone = Brown
Giallo = Yellow
Azzurro = Blue
Rosso/Nero = Red/Black
Azzurro/Nero = Blue/Black
Verde/Nero = Green/Black
Bianco/Nero = White/Black
Giallo/Nero = Yellow/Black
Grigio/Nero = Grey/Black
Grigio/Rosso = Grey/Red

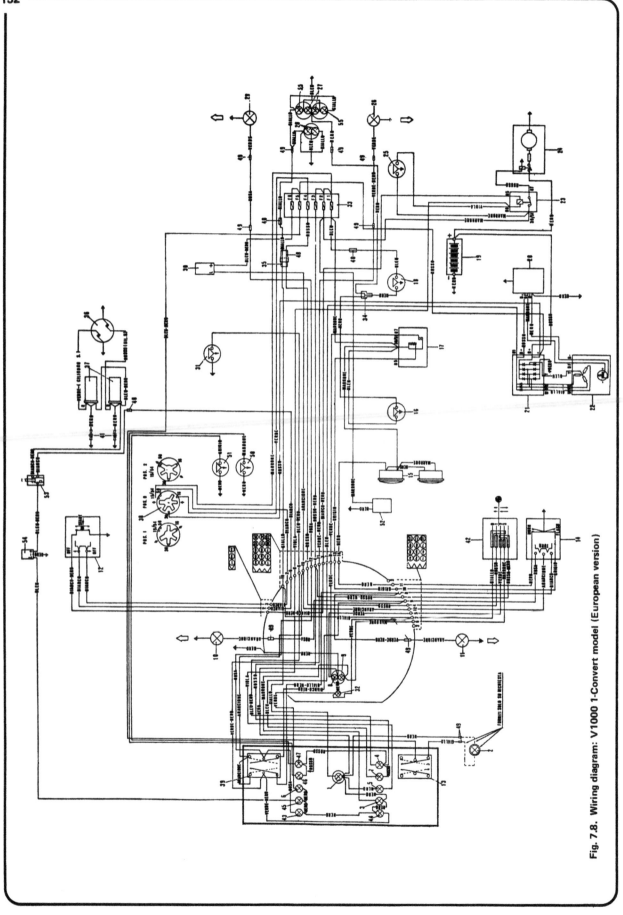

Fig. 7.8. Wiring diagram: V1000 1-Convert model (European version)

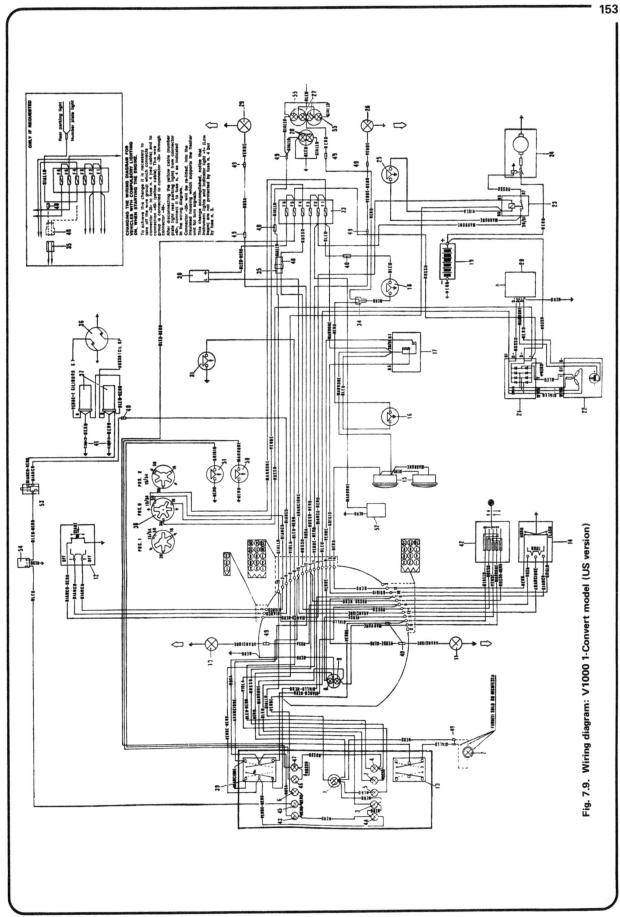

Fig. 7.9. Wiring diagram: V1000 1-Convert model (US version)

**Colour code**

| | | |
|---|---|---|
| Nero | = | Black |
| Bianco | = | White |
| Verde | = | Green |
| Grigio | = | Grey |
| Viola | = | Violet |
| Arancio | = | Orange |
| Rosa | = | Pink |
| Rosso | = | Red |
| Marrone | = | Brown |
| Giallo | = | Yellow |
| Azzurro | = | Blue |
| Rosso/Nero | = | Red/Black |
| Azzurro/Nero | = | Blue/Black |
| Verde/Nero | = | Green/Black |
| Bianco/Nero | = | White/Black |
| Giallo/Nero | = | Yellow/Black |
| Grigio/Nero | = | Grey/Black |
| Grigio/Rosso | = | Grey/Red |

1 Mile counter, speedometer (bulb 3W)
2 Additional light (bulb 5) - only on request
3 High beam warning light (bulb 1.2W) H
4 Oil pressure warning light (bulb 1.2W) Oil
5 Neutral position warning light (bulb 1.2W) N
6 Low beam and parking warning light (bulb 1.2W) L
7 Generator charge warning light (bulb 1.2W) Gen
8 Low beam    bulb 40/45W
9 High beam
10 Turn indicator light - right, front (bulb 21W)
11 Turn indicator light - left, front (bulb 21W)
12 Engine starting and stopping control
13 Additional light switch
14 Control: turn indicator lights, horns, flashing lights
15 Horns (consumption: 7A)
16 Front brake switch
17 Flashing light (flash) relay
18 Rear brake switch
19 Battery
20 Regulator
21 Rectifier
22 Alternator
23 Starter motor relay
24 Starter motor
25 Switch on clutch control wire
26 Turn indicator light - rear, left (bulb 21W)
27 Rear stop light (bulbs 5/21W)
28 Number plate light (bulb 5W)
29 Turn indicator lights, flasher unit

30 Turn indicator lights, flasher unit
31 Oil pressure switch (on the engine crankcase)
32 Town driving light, front (bulb 3W)
33 Terminal block with fuses (16A fuses)
34 3-way connector
35 4-way connector (amp)
36 Breaker
37 Coils
38 Ignition switch (3 positions).
39 Switch actuating rear turn indicator lights flashing
40 2-way connector
41 Spark plugs
42 Light switch with travel limit from position
43 High/Low beam to position  Town driving light
    Right turn indicator warning light (bulb 1.2W)
44 Left turn indicator warning light (bulb 1.2W)
45 Warning light indicating  side stand  in position
    Park (bulb 1.2W)
46 Brake fluid level warning light  Brake (bulb 1.2W)
47 Fuel level warning light  Fuel (bulb 1.2W)
48 4-way connector (amp)
49 Connection
50 Brake fluid level indicator
51 Fuel level indicator
52 Electrovalve (2.5W)
53 Coil control device
54 Commutator for side stand warning light
    Park position
55 Rear parking light (bulb 5/21W)

Fig. 7.9. Wiring diagram: V1000 1 - Convert model (US version)

# Metric conversion tables

| Inches | Decimals | Millimetres | Millimetres to Inches | | Inches to Millimetres | |
|--------|----------|-------------|-----|--------|--------|----|
| | | | mm | Inches | Inches | mm |
| 1/64 | 0.015625 | 0.3969 | 0.01 | 0.00039 | 0.001 | 0.0254 |
| 1/32 | 0.03125 | 0.7937 | 0.02 | 0.00079 | 0.002 | 0.0508 |
| 3/64 | 0.046875 | 1.1906 | 0.03 | 0.00118 | 0.003 | 0.0762 |
| 1/16 | 0.0625 | 1.5875 | 0.04 | 0.00157 | 0.004 | 0.1016 |
| 5/64 | 0.078125 | 1.9844 | 0.05 | 0.00197 | 0.005 | 0.1270 |
| 3/32 | 0.09375 | 2.3812 | 0.06 | 0.00236 | 0.006 | 0.1524 |
| 7/64 | 0.109375 | 2.7781 | 0.07 | 0.00276 | 0.007 | 0.1778 |
| 1/8 | 0.125 | 3.1750 | 0.08 | 0.00315 | 0.008 | 0.2032 |
| 9/64 | 0.140625 | 3.5719 | 0.09 | 0.00354 | 0.009 | 0.2286 |
| 5/32 | 0.15625 | 3.9687 | 0.1 | 0.00394 | 0.01 | 0.254 |
| 11/64 | 0.171875 | 4.3656 | 0.2 | 0.00787 | 0.02 | 0.508 |
| 3/16 | 0.1875 | 4.7625 | 0.3 | 0.1181 | 0.03 | 0.762 |
| 13/64 | 0.203125 | 5.1594 | 0.4 | 0.01575 | 0.04 | 1.016 |
| 7/32 | 0.21875 | 5.5562 | 0.5 | 0.01969 | 0.05 | 1.270 |
| 15/64 | 0.234275 | 5.9531 | 0.6 | 0.02362 | 0.06 | 1.524 |
| 1/4 | 0.25 | 6.3500 | 0.7 | 0.02756 | 0.07 | 1.778 |
| 17/64 | 0.265625 | 6.7469 | 0.8 | 0.3150 | 0.08 | 2.032 |
| 9/32 | 0.28125 | 7.1437 | 0.9 | 0.03543 | 0.09 | 2.286 |
| 19/64 | 0.296875 | 7.5406 | 1 | 0.03937 | 0.1 | 2.54 |
| 5/16 | 0.3125 | 7.9375 | 2 | 0.07874 | 0.2 | 5.08 |
| 21/64 | 0.328125 | 8.3344 | 3 | 0.11811 | 0.3 | 7.62 |
| 11/32 | 0.34375 | 8.7312 | 4 | 0.15748 | 0.4 | 10.16 |
| 23/64 | 0.359375 | 9.1281 | 5 | 0.19685 | 0.5 | 12.70 |
| 3/8 | 0.375 | 9.5250 | 6 | 0.23622 | 0.6 | 15.24 |
| 25/64 | 0.390625 | 9.9219 | 7 | 0.27559 | 0.7 | 17.78 |
| 13/32 | 0.40625 | 10.3187 | 8 | 0.31496 | 0.8 | 20.32 |
| 27/64 | 0.421875 | 10.7156 | 9 | 0.35433 | 0.9 | 22.86 |
| 7/16 | 0.4375 | 11.1125 | 10 | 0.39270 | 1 | 25.4 |
| 29/64 | 0.453125 | 11.5094 | 11 | 0.43307 | 2 | 50.8 |
| 15/32 | 0.46875 | 11.9062 | 12 | 0.47244 | 3 | 76.2 |
| 31/64 | 0.484375 | 12.3031 | 13 | 0.51181 | 4 | 101.6 |
| 1/2 | 0.5 | 12.7000 | 14 | 0.55118 | 5 | 127.0 |
| 33/64 | 0.515625 | 13.0969 | 15 | 0.59055 | 6 | 152.4 |
| 17/32 | 0.53125 | 13.4937 | 16 | 0.62992 | 7 | 177.8 |
| 35/64 | 0.546875 | 13.8906 | 17 | 0.66929 | 8 | 203.2 |
| 9/16 | 0.5625 | 14.2875 | 18 | 0.70866 | 9 | 228.6 |
| 37/64 | 0.578125 | 14.6844 | 19 | 0.74803 | 10 | 254.0 |
| 19/32 | 0.59375 | 15.0812 | 20 | 0.78740 | 11 | 279.4 |
| 39/64 | 0.609375 | 15.4781 | 21 | 0.82677 | 12 | 304.8 |
| 5/8 | 0.625 | 15.8750 | 22 | 0.86614 | 13 | 330.2 |
| 41/64 | 0.640625 | 16.2719 | 23 | 0.90551 | 14 | 355.6 |
| 21/32 | 0.65625 | 16.6687 | 24 | 0.94488 | 15 | 381.0 |
| 43/64 | 0.671875 | 17.0656 | 25 | 0.98425 | 16 | 406.4 |
| 11/16 | 0.6875 | 17.4625 | 26 | 1.02362 | 17 | 431.8 |
| 45/64 | 0.703125 | 17.8594 | 27 | 1.06299 | 18 | 457.2 |
| 23/32 | 0.71875 | 18.2562 | 28 | 1.10236 | 19 | 482.6 |
| 47/64 | 0.734375 | 18.6531 | 29 | 1.14173 | 20 | 508.0 |
| 3/4 | 0.75 | 19.0500 | 30 | 1.18110 | 21 | 533.4 |
| 49/64 | 0.765625 | 19.4469 | 31 | 1.22047 | 22 | 558.8 |
| 25/32 | 0.78125 | 19.8437 | 32 | 1.25984 | 23 | 584.2 |
| 51/64 | 0.796875 | 20.2406 | 33 | 1.29921 | 24 | 609.6 |
| 13/16 | 0.8125 | 20.6375 | 34 | 1.33858 | 25 | 635.0 |
| 53/64 | 0.828125 | 21.0344 | 35 | 1.37795 | 26 | 660.4 |
| 27/32 | 0.84375 | 21.4312 | 36 | 1.41732 | 27 | 685.8 |
| 55/64 | 0.859375 | 21.8281 | 37 | 1.4567 | 28 | 711.2 |
| 7/8 | 0.875 | 22.2250 | 38 | 1.4961 | 29 | 736.6 |
| 57/64 | 0.890625 | 22.6219 | 39 | 1.5354 | 30 | 762.0 |
| 29/32 | 0.90625 | 23.0187 | 40 | 1.5748 | 31 | 787.4 |
| 59/64 | 0.921875 | 23.4156 | 41 | 1.6142 | 32 | 812.8 |
| 15/16 | 0.9375 | 23.8125 | 42 | 1.6535 | 33 | 838.2 |
| 61/64 | 0.953125 | 24.2094 | 43 | 1.6929 | 34 | 863.6 |
| 31/32 | 0.96875 | 24.6062 | 44 | 1.7323 | 35 | 889.0 |
| 63/64 | 0.984375 | 25.0031 | 45 | 1.7717 | 46 | 914.4 |

# English/American terminology

Because this book has been written in England, British English component names, phrases and spellings have been used throughout. American English usage is quite often different and whereas normally no confusion should occur, a list of equivalent terminology is given below.

| English | American | English | American |
|---------|----------|---------|----------|
| Air filter | Air cleaner | Number plate | License plate |
| Alignment (headlamp) | Aim | Output or layshaft | Countershaft |
| Allen screw/key | Socket screw/wrench | Panniers | Side cases |
| Anticlockwise | Counterclockwise | Paraffin | Kerosene |
| Bottom/top gear | Low/high gear | Petrol | Gasoline |
| Bottom/top yoke | Bottom/top triple clamp | Petrol/fuel tank | Gas tank |
| Bush | Bushing | Pinking | Pinging |
| Carburettor | Carburetor | Rear suspension unit | Rear shock absorber |
| Catch | Latch | Rocker cover | Valve cover |
| Circlip | Snap ring | Selector | Shifter |
| Clutch drum | Clutch housing | Self-locking pliers | Vise-grips |
| Dip switch | Dimmer switch | Side or parking lamp | Parking or auxiliary light |
| Disulphide | Disulfide | Side or prop stand | Kick stand |
| Dynamo | DC generator | Silencer | Muffler |
| Earth | Ground | Spanner | Wrench |
| End float | End play | Split pin | Cotter pin |
| Engineer's blue | Machinist's dye | Stanchion | Tube |
| Exhaust pipe | Header | Sulphuric | Sulfuric |
| Fault diagnosis | Trouble shooting | Sump | Oil pan |
| Float chamber | Float bowl | Swinging arm | Swingarm |
| Footrest | Footpeg | Tab washer | Lock washer |
| Fuel/petrol tap | Petcock | Top box | Trunk |
| Gaiter | Boot | Torch | Flashlight |
| Gearbox | Transmission | Two/four stroke | Two/four cycle |
| Gearchange | Shift | Tyre | Tire |
| Gudgeon pin | Wrist/piston pin | Valve collar | Valve retainer |
| Indicator | Turn signal | Valve collets | Valve cotters |
| Inlet | Intake | Vice | Vise |
| Input shaft or mainshaft | Mainshaft | Wheel spindle | Axle |
| Kickstart | Kickstarter | White spirit | Stoddard solvent |
| Lower leg | Slider | Windscreen | Windshield |
| Mudguard | Fender | | |

# Index